Man, the Chemical Machine

Man

the Chemical Machine

by ERNEST BOREK

Columbia University Press, New York, 1952

For my Sisters

Elizabeth and *Irene*

Preface

BIOLOGICAL CHEMISTRY, the study of the chemical structure and chemical mechanism of living things, is at the present the most rapidly growing, and in some ways, the most spectacular of the biological sciences. Yet, oddly, the field has remained almost hidden from the general public. To be sure, the troubadours of the Sunday supplements have sung the glories of some of the achievements of the biochemist. But such latter day *chansons de geste* are invariably preoccupied with the spectacular clinical applications of biological chemistry. (And, incidentally, they habitually credit the wrong hero, modern medicine instead of modern chemistry, with the brilliant deeds.)

As far as I know there has not appeared as yet a connected story showing the biochemist at his bench, tracing the growth of the ideas that guide his hands and unfolding his view on the mechanism of the living machine. It is such a story of my field that I tried to tell in this book. Therefore I set out to emphasize not the final clinical applications of the biochemist's efforts but rather the accumulation of ideas, bold deductions, and sometimes, fruitful accidents which led to those medical bounties.

To reach a larger audience I decided to dispense with chemical formulas completely. Those formulas are merely shorthand symbols for facts and ideas. The formula H-O-H merely states that a water molecule is made up of two atoms of the element hydrogen and one of the element oxygen and it also tells something of the energy binding the atoms together. It no more depicts the exact appearance of a water molecule than a blueprint depicts the exact appearance of the engine which will be built from it. In both cases the symbols have complete meaning only for the technician but they can be translated to any intelligent lay person. An apt word is worth more than a thousand obscure pictures.

While describing the work of the biochemist without using his baffling symbols I tried to follow the precept for the exposition of a science to a lay audience given by the great physicist James Clerk Maxwell: "For the sake of persons of different types, scientific truth should be presented in different forms and should be regarded as equally scientific, whether it appears in the robust form and vivid coloring of a physical illustration, or in the tenuity and paleness of a symbolical expression."

I should emphasize that while I made every effort to achieve "scientific truth," I did not attempt to present the development of the various areas of the biochemist's endeavor in complete historical sequence. I felt that an approach which would undoubtedly yield a very long and tedious book could not be risked in a work which is aimed not at the captive audience of students but rather at the free public. In most of the chapters the historical background was sketched in only to serve as a frame into which current concepts could be woven. In turn, only such current biochemical work was included which could be fitted into a pattern within the frame.

I am not an admirer of the genre of science reporting in which the work of the scientist is but a thin filling sandwiched between soggy chunks of elaborate accounts of the real and imaginary drama of his life. ("A haggard man in white is pacing in the dim laboratory. His eyes, granules of shining coal in dark pits; his mouth, a thin, tightly drawn ribbon as he mutters: 'This time it must work!' ".) Even if I were an admirer of this sort of thing I could not perpetrate it in this book. Biological chemistry is a young science; many of the scientists whose work is included in the book are very much alive. Indeed, I concentrated on the science rather than on the scientist to the point where I did not even identify by name many of the contemporary biochemists—often only their work is mentioned. I thus managed to evade the responsibility of choosing whom to include from among fellow scientists many of whom are acquaintances and some personal friends.

Biological chemistry is developing so rapidly that even the textbooks do not achieve a synthesis of the field but merely compile the emerging facts. I therefore appealed to several experts for help to weed out errors from this attempt at a synthesis. I list these colleagues not to hide behind them if errors should be pointed out, but to thank them for their cordial help:

Drs. Elvin Kabat, Irving London, David Nachmansohn, David Shemin, David Rittenberg, Francis J. Ryan, Salome Waelsch, and Heinrich Waelsch, all of Columbia University, and Dr. Abraham Mazur of City College and Cornell University Medical School.

I am also indebted to my wife Blanche Ann Borek for scientific help and personal encouragement.

I had final assurance to proceed with the publication of the book after Dr. H. T. Clarke, a past president of the American Society of Biological Chemists and a master stylist of scientific prose, kindly read and criticized the whole book.

The great explorer of the Himalayas, the late F. S. Smyth, wrote that in a chronicle of exploration what is interesting is not merely the reaching of the goal but the striving for it: "The kingdom of adventure, not the crown." I hope the reader will accept these pages as a guidebook to the kingdom of biochemistry.

ERNEST BOREK

New York
May 21, 1952

Acknowledgments

I AM INDEBTED to the John Simon Guggenheim Memorial Foundation for a Fellowship during whose tenure I put the book into final shape and also to the Staff of the Columbia University Press for believing with me that the biochemist's story is worth presenting to the public and for their patient tutoring in the rites of bookmaking.

Longmans, Green & Co. kindly gave permission to quote from Pierre Lecomte du Noüy's *The Road to Reason*. Excerpts from Pasteur's acceptance speech to the Academy were taken from *The Life of Pasteur* by René Vallery-Radot.

The illustration at the head of Chapter 4 is reproduced from "The Chemistry of Glycogen," by Kurt H. Meyer, in *Advances in Enzymology*, Vol. III (1943), F. F. Nord, ed., by permission of Interscience Publishers, Inc., New York–London.

The illustration at the head of Chapter 7 is reproduced from R. W. Hegner, *College Zoology*, page 393 (5th ed., 1942) by permission of the Macmillan Company.

An article "The Cornerstone of our Tissues," based on Chapter 6, has been acquired by the magazine *Your Health*.

E. B.

Contents

Man, the Chemical Machine

1 The Stuff of Life

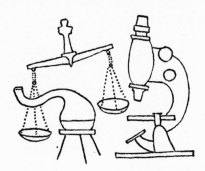

A FERTILIZED EGG is at once the most precious and the most baffling thing in the universe. Within the thin shell of a hen's egg is locked not just a mere mass of egg white and yolk, but a promise of a beautiful creature of flesh, blood, and bones—a promise of the continuity of life.

The mode of fulfillment of that promise is the most baffling mystery within the horizon of the human mind. The mystery of the sun's energy no longer eludes us: we have converted mass into energy; we have unleashed minuscule suns over Hiroshima and Nagasaki. But the mastery of the tools, of the energy, and of the blueprint which shapes an apparently inert mass into life, is still but a thin hope.

A century and a quarter ago there was not even such a hope. In our striving for knowledge we had to contend not only with the secret ways of the universe, but often we had to surmount man-made barriers before we could approach our quest. One such road block in the path to knowledge was the principle of vitalism which dominated scientific thought until the middle of the nineteenth century. Scientists had explored with fruitful zest the nonliving world, but they halted with awe and impotence before a living thing. It was believed by the vitalists that the cell membrane shrouded mysterious vital forces and "sensitive spirits." It was an unassailable tenet that not only could we not fathom these mysteries, but that we should never be able to duplicate by any method a single product of such vital forces. An uncrossable chasm was supposed to separate the realm of the living, organic world and the realm of the nonliving, inorganic world.

In 1828, a young man of twenty-eight unwittingly bridged that chasm. He made something in the chemical laboratory which, until then, had been made only in the body of a living thing. This achievement was the "atom bomb" of the nineteenth century. Its influence in shaping our lives is far greater than the influence of atomic energy will be.

Had there been science writers on the newspapers of that day they could very well have unfurled all the clichés of their present-day counterparts about the achievement. But, oddly, not only were the man and his feat unknown to his contemporaries, he is practically unknown even today. The generals of the Napoleonic era—Blucher, Ney, Bernadotte—are known to many, but their contemporary, Friedrich Wöhler, who was a giant of intellect and influence compared to them, is known only to chemists.

Friedrich Wöhler was a student of medicine at Heidelberg in the early 1820s. His chemistry teacher, Gmelin, was one of those rare teachers who not only dispensed knowledge but "shaped souls." Under his guidance Wöhler left medicine and became a chemist. He more than justified his teacher's faith, for, in addition to the great discovery which shook the foundation of vitalism, we owe to Wöhler the discovery of the two elements aluminum and beryllium. After absorbing all that Heidelberg could offer, he went to Sweden to work with Berzelius, the greatest contemporary master of chemistry.

There Wöhler discovered, quite by accident, a method of making urea, a substance theretofore produced only by the cells of living creatures. So contrary to current ideas was this achievement that he published it only after numerous repetitions, four years later. How did he learn the secret of the sensitive spirits? How did he make urea?

Urea is a substance found as a waste product in the urine of some animals. It can be cajoled out in the form of pure white crystals, by the knowing hands of the chemist. Like all other pure compounds, urea has characteristic attributes which distinguish it from any other substance. Sugar and salt are two different pure compounds which superficially look alike. But the tongue tells them apart with unfailing ease. Their different impact on our discerning little taste buds is but one of many differences between salt and sugar. The chemist has discovered scores of other differences. The elements of which they are composed, the temperature at which they melt, certain optical properties of the crystals, these are some of the distinguishing lines in the fingerprint of a compound, by means of which the chemist can recognize a particular one among the multitudes.

Urea happens to be made of four different elements. One atom of carbon, one of oxygen, two of nitrogen, and four of hydrogen make up a urea molecule. These atoms are attached to each other in a definite pattern, a pattern unique to a single substance, urea. The force which binds these atoms into the pattern of urea is the energy in the electrons of those atoms. A chemical union between atoms is a light, superficial affair. Two atoms meet, some of their outer electrons become entangled and a temporary union is formed. The nucleus of the atom is completely unaffected by a chemical union. The energy involved in the making or breaking of such a union is minuscule compared to the vast energy locked within the nucleus of the atom—the monstrous nuclear, or atomic, energy.

How do we know that there are four hydrogens, two nitrogens, one carbon, and one oxygen in urea? Furthermore, how do we know the pattern they form? The writer begs the reader's indulgence for not answering these questions. Scientists often accept the conclusions of another science merely to save time; a geologist studying the movement of glaciers accepts the density of ice given to him by the physicist. In this age of specialization the structure of molecules does not lie within the biochemist's sphere of responsibility. The deciphering of the architecture of molecules is assigned to other specialists, the physical chemist and the organic chemist. The biochemist accepts their conclusions and it is hoped that the reader will, too.

Until Wöhler succeeded in making it, urea could be fashioned only in a living animal by unknown, awesome, animated spirits. The spirits whimsically flung out their product into the urine for reasons, it was thought, no human could fathom.

Wöhler had no intention of making urea in a test tube. He was studying a simple, undramatic, chemical reaction. He wanted to make a new inorganic compound, ammonium cyanate. He went through a variety of manipulations which were expected to yield the new substance. As the final step, he boiled away the water and some white crystals were left behind. But they were not the new inorganic salt that he had expected; they were the very same urea which animals excrete!

It so happens that in ammonium cyanate, the substance Wöhler had set out to make, there are the same elements in the same number as in urea. The difference is the pattern the atoms form. The pattern of ammonium cyanate was disrupted by the heat of the boiling water and the atoms rearranged themselves to form urea. (The changing of chemical structures by heat is not unusual; indeed it is an everyday household feat—a boiled egg is quite different from an uncooked one.)

Simple? It looks simple now, 120 years later, when animated spirits have become scientific antiques, and the manufacture of synthetic vitamins is a big industry. Thus, only in our hindsight, which sharpens with the years elapsed, does it look simple.

The importance of the finding was not lost to Wöhler or his contemporaries. While he wrote very modestly of his achievement in his four-page technical communication, he let himself go when writing to his mentor, Berzelius. "I must now tell you," he wrote, "that I can make urea without calling on my kidneys, and indeed, without the aid of any animal, be it man or dog."

And the master replied graciously: "Like precious gems—aluminum and artificial urea—two very different things coming so close together [1]—have been woven into your laurel wreath."

[1] Wöhler discovered the new element aluminum in 1827 and announced the synthesis of urea in 1828.

The homage of history was paid by Sir Frederick Gowland Hopkins, Nobel Prize winner, on the centenary of the discovery: "The intrinsic historic importance of Wöhler's synthesis can hardly be exaggerated. So long as the belief held ground that substances formed in the plant or animal could never be made in the laboratory, there could be no encouragement for those who instinctively hoped that chemistry might join hands with biology. The very outermost defences of vitalism seemed unassailable."

Though the barrier between organic and inorganic chemistry was demolished by Wöhler, the terms have been retained, but with new meanings. Organic chemistry now embraces the chemistry of the compounds of carbon. This element is uniquely fecund in forming compounds. Close to half a million different compounds of carbon have been made by organic chemists. And new ones, by the dozen, are being added daily. The compounds prepared prior to 1930 are briefly described in a "handbook" of some sixty volumes, of about a thousand pages each. In those pages are hidden veritable mines of drugs, perfumes, and plastics. It is impossible to predict what uses a new compound may have. Sulfanilamide, DDT, four members of the vitamin B group, and many other drugs lay for years on the shelf of organic chemistry before their potency, in nutrition and medicine, was discovered.

To inorganic chemistry is relegated the study of the compounds of the other elements. All of these together number only a paltry 25,000.

Encouraged by Wöhler's success, chemists threw themselves on the problem of the products and the constituents of the living cell. With ravenous zest they started to tear cells apart, and today, 120 years later, they still have not been sated; they are still at it. Hundreds of compounds which had been pried out from the cell were duplicated [2] in the laboratory after Wöhler's fashion.

These successes changed medicine, changed nutrition, changed our way of life. Are you eating vitamin-enriched bread? Were you given large doses of vitamin K before an operation to stop excessive hemorrhage? Are you receiving injections of hormones? Has the life of a dear one been saved by penicillin? For all these bounties, thank Wöhler and the generations of chemists his deed encouraged.

Nor was the reproduction of some of the cell's constituents completely satisfying to the chemists and other scientists. They were emboldened. If we can duplicate some of the products of the cell maybe we can observe and understand the cell's vitalistic mechanisms with which *it* makes these products.

So, haltingly, they started to study what takes place in the cell, what converts the egg into a chick. To their utter amazement and delight they found, again, not mysterious vitalistic mechanisms but a whole series of earth-bound chemical reactions. But what an astounding number of different reactions! And on what a minute scale these reactions take place! And in what well-ordered sequence!

Just one example of the cell's astonishing schedule of operation: that only one child out of a thousand is born with a hare lip is testimony to the precision in the development of the other nine hundred and ninety-nine. The upper

[2] The term synthetic has acquired an opprobrious meaning, indicating a poor substitute for the genuine. However, when the chemist synthesizes a known substance he makes an identical duplicate of what nature had made. Vitamin C, whether extracted from an orange, or from the vat of the chemical manufacturer, is exactly the same material; it is impossible to tell the products apart by any means.

lip in the human embryo is formed at the end of the second month of gestation. To form it, a little bud of cells begins to grow downward from the nose. At the same time two other little buds from the left and right cheek begin to grow toward the center. The three minute pieces meet at the same time and fuse; the two ridges running from our nose to our lips are the seams left by this fusion. If one of the buds is too slow in making its appointment, the edge of the other waiting piece loses its ability to fuse and the child is born with a malformed lip. (Fortunately the hand of a skilled surgeon can remedy this rare blunder of laggard cells.)

However complex the reactions in the cell, on whatever minute scale, they are chemical mechanisms. Fragments of the cell's chemical machinery even when extracted from the cell continue functioning, just like any other well-behaved chemical reaction.

Gone is the paralyzing awe with which pre-Wöhler scientists beheld a living thing. Gone is the defeatist conviction that activities in the cell can not be revealed to man.

Scientists use a variety of techniques and approaches in the study of the cell. Those who are primarily interested in the chemical make-up and chemical mechanisms are the biological chemists. They were always a small band—the American Society of Biological Chemists has fewer than a thousand members—but they have achieved much since Wöhler's time.

In this chronicle of the achievements of those chemists we may be carried away by our enthusiasm and pride in our ever-growing prowess and knowledge. We should at such times recall what the discoverer of the circulation of blood said in 1625. "All we know," said William Harvey,

"is still infinitely less than all that still remains unknown."
That humble statement is as true today as it was then, for
325 years is a very short time to study anything as wonderfully complex as a living cell.

What have the generations of biochemists found in living things? First of all, they struck water. Lots of water.
About 70 percent of the human body is water. "Water thou
art and to water returnest" would be a chemically more accurate, if less euphonious, description of our corporal de-
nouement. The amount of water in the human body is sur-
prisingly constant. When it increases locally in a small area
the tissues become swollen. There is a general increase in
the water content of tissues in old age. The shrunken, ex-
ternally parched appearance of old age is misleading, for
the water content of the body is actually increased.
Whether this increased hydration has any causal relation
to aging is one of the multitude of unanswered questions
which makes Harvey's humble statement all too true.

We must not find in the hydration of old age justification
for replacing water as a staple beverage by more potent
fluids. Alcohol actually introduces more water into the body
than a similar weight of water does. An ounce of absolute,
two-hundred-proof alcohol, will produce about one and one
sixth ounces of water. The formation of that much water,
although surprising, is perfectly possible. One of the constituent elements of alcohol is hydrogen. When alcohol is
burned in the body the hydrogen is combined with oxygen
that we inhale from the atmosphere, to form water.

This type of water formation is not unique to alcohol;
every food is a source of water in a living creature. The
camel puts this bit of biochemistry to a very practical use.

To cross the desert he needs both food and water. The camel's hump, which is largely fat, provides both. Assuming that a camel's hump contains 100 pounds of fat, the camel will derive from burning that fat huge amounts of energy—over 400,000 Calories—and as a bonus, 50 quarts of water.

Water serves us well: it is a freight canal for the transport of foods and wastes; it regulates the body temperature by evaporating as cooling is needed; it is the remarkably efficient lubricating fluid for the body's many joints; and, finally, it makes up 70 percent of the human body. We living creatures contain only 30 percent solids. We have about the same proportion of solids as a cup of water containing ten teaspoonfuls of sugar. Why don't we flow as freely as that sugar solution does? Why is our flesh, in Hamlet's words, "too, too solid"? Why doesn't it "melt, thaw and resolve itself into a dew"?

The biochemist's prosaic answer to the prince would be: "Because we have proteins."

Proteins, which are the most challenging and least-known constituents of the cell, are able to bind large amounts of water into themselves, forming semisolid jellies. Anyone who has ever made gelatin dessert knows this. A small amount of a dry protein—gelatin—soaks up a large volume of water and produces a semirigid mass. Not all of our proteins bind water as readily as gelatin does; we must have considerable amounts of free, unbound water in our bodies. But such water is usually confined within tubes or tissues and thus we can retain the body's characteristic solidity.

Proteins are the most characteristic components of the cell; all of life's processes are tied up in them. Indeed, there are proteins—the viruses—which have many of the attributes of life. (The term protein has been very aptly devised. It means, of primary importance.)

We find proteins in every cell. Egg white, cheese, hair, and nails are composed largely of proteins. If a protein is boiled with strong acid for twenty hours it loses its identity and its characteristic properties. The edifice of the protein molecule is crumbled into its component bricks by the hot acid. The bricks that we can find in a solution of a dismantled protein are the amino acids.

There are more than twenty different amino acids. They all contain the element nitrogen. Hence the great need for nitrogenous fertilizers to insure good crops. Plants need the nitrogen from the soil to fashion their amino acids and proteins.

The various amino acids differ in their chemical patterns. However, a small part of their structure, the part by means of which they are joined together, is repeated in each one. They each have, as it were, an identical "hook" and an "eye" for ease of joining, but otherwise they are different.

We owe to a great German chemist, Emil Fischer, our knowledge of the way in which amino acids are linked together in a protein molecule. To achieve a union, two amino acids together shed off a molecule of water. Part of the water molecule comes from the "hook" of one amino acid and part from the "eye" of the other. The electronic forces which originally had held onto the atoms which form the water now graft the two amino acids together. In other words, an atomic fragment is carved out of each amino acid and fusion takes place at the shorn sites. (The carved-out fragments form a molecule of water.) Only the "eye" of one amino acid and the "hook" of the other have been engaged to join the two together. But each amino acid has both an "eye" and a "hook." Therefore one "eye" and one "hook" are left unengaged in this couplet of amino acids. By means of these unemployed coupling devices two more amino acids can be attached to form a unit of four. This accretion can continue until hundreds of amino acids are linked to form the huge chain which is a protein molecule.

Proteins are the largest molecules known. Some of them are several million times heavier than the hydrogen atom, but even these are submicroscopic in size.

The sequence in which the amino acids are fused together is a mystery which still baffles us and will continue to do so for a long time to come. Variations in the sequence of adjoining amino acids can produce a vast variety of different proteins. Imagine a thousand children from each of twenty different nations, dressed in their national costume and joined into a huge daisy chain. If they overcome their nationalistic antipathies and intermingle, we have a crude analogy between them and twenty different amino acids, repeated a thousand times each, to form a protein molecule. The number of different chains of children—or of amino acids—that can be formed, runs into millions.

We can now appreciate why the structure of proteins presents staggering difficulties to the biochemist. The proteins are the cell's last stronghold which continues to defy the prying efforts of Wöhler's successors. Since many of the cell's pivotal mechanisms are intimately linked to the structure of proteins, our ignorance of the latter is a tremendous handicap. However, as other, previously baffling obstacles have yielded, this too, will eventually succumb to the chemist's fruitful curiosity.

What else, besides water and proteins, does the biochemist find in the cell? He finds fats and sugars which, along with protein, make up the solid matter of the cell. (The fats and sugars will be subjected to chemical scrutiny in later chapters.)

In addition to water, proteins, fats, and sugars, the cell contains, in minute amounts, scores of other organic substances, such as vitamins and hormones. It also contains a large number of inorganic salts. Some of the salts are present in considerable amounts; of others we find but traces.

Water, proteins, fats, sugars, and salts, such are the mundane substances the chemist finds within a living cell. Certainly these are not "stuff as dreams are made of." And yet, of such stuff is built the edifice of the improbable dream that is life. All the greater, therefore, is the miracle of that life.

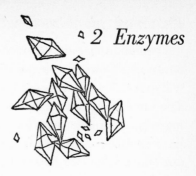

Giant molecules with gigantic know-how

WE LIVE because we have enzymes. Everything we do—walking, thinking, reading these lines—is done with some enzyme process. A prominent enzyme chemist defined life as "a system of cooperating enzyme reactions."

The best way to get acquainted with an enzyme is to observe it in action. In order to break down a protein, say some egg white, into its constituent amino acids in the laboratory, we must use rather drastic methods. We add ten times its weight of concentrated acid and boil the mixture for twenty hours. In the stomach and small intestine the same disintegration of the egg takes place in a couple of hours at body temperature and without such strong acid. This chemical sleight of hand is performed by the enzymes made for this purpose by the cells in the stomach wall. If we add the stomach lining of a recently slaughtered hog to boiled egg white and keep the mixture at body temperature for a few hours, the egg white will be disintegrated into its amino acids just as effectively as it would have been in the stomach of the live animal.

Their enormous activity is characteristic of enzymes. One ounce of an enzyme preparation from the hog's stomach will digest 50,000 ounces of boiled egg white in two hours. The same preparation will also clot milk—it is the active ingredient of rennet powders. The potency of enzymes can be demonstrated even more impressively with this activity: one ounce will clot 2,800,000 quarts of milk!

It is astonishing how recent is our knowledge of biological mechanisms. Until the early 1820s we had no conception of what happens to the food in our mysterious interiors. Prior to that, there were a host of conjectures, not the least imaginative of which was that a band of little demons are busily engaged in our stomachs, macerating our food.

Toward the end of the eighteenth century Réaumur in France and Spallanzani in Italy took the first steps toward the exploration of the stomachs of animals. They fed food, enclosed in a wire cage or in a perforated capsule, to animals and, at various intervals, retrieved the containers by means of attached strings. They noted the dissolution of the food but could not even guess at the nature of the substances which were responsible for these changes.

A whole set of new explanations was brought forth, but that they did not gain universal acceptance is obvious from the irascible fragment of a lecture by the English physician William Hunter: "Some physiologists will have it, that the stomach is a mill, others, that it is a fermenting vat, others, again that it is a stew pan; but in my view of the matter, it is neither a mill, a fermenting vat, nor a stew pan; but a stomach, gentlemen, a stomach."

An accident in 1822 literally lifted the veil which covered the human stomach and its disputed processes. Fortunately, a man of rare ability was on hand to exploit this opportunity. Dr. William Beaumont was a surgeon in the recently organized U.S. Army Medical Corps. He was in charge of the post hospital at Fort Mackinac, on an island in northern Lake Michigan. The island was a busy center of fur trading. Trappers and voyageurs would swarm toward this post, their canoes laden with the winter's yield of fur pelts.

One such French Canadian voyageur, Alexis St. Martin, was part of the usual crowd at the trading post of the American Fur Company on June 6, 1822. Apparently by accident, someone's gun was discharged and St. Martin received the whole load, at short range in his abdomen. The young army surgeon came on the double to the aid of the victim. On arrival, however, he realized that he could be of but little help. The bullets had ripped a huge wound, through which were protruding large chunks of lungs and what appeared to be ripped pieces of the stomach. Beaumont dressed the wound as best he could, cutting away some of the protruding flesh with a penknife, and had the patient carried to the shack which served as the hospital.

Had Beaumont had any colleagues on the post, he undoubtedly would have described the wound, between gulps of his supper, and would have expressed the doubts he had of St. Martin's chances of recovery. But Beaumont was all alone on this and on subsequent posts. His achievements are, therefore, all the more remarkable.

To Beaumont's amazement the sturdy youth survived the night. There followed a heroic struggle by Beaumont to save the patient; operation followed operation; no effort was spared to dress and drain the slowly healing wound. After several months the town officials refused to support the convalescent any longer; they were ready to ship him back where he came from—Canada—by open boat. Motivated by both charity and interest, Beaumont took the youth to his own home where he continued to nurse and observe him.

For on St. Martin there was something to observe which had, undoubtedly, never before been seen by the human eye; an open window into a normally functioning human stomach! The gaping wound in the stomach never sealed. Its edges became healed, but a large hole in the upper part of the stomach remained open for the rest of St. Martin's unexpectedly long life. (He died at the age of 83.)

Beaumont recognized his great opportunity, for he could "look directly into its [the stomach's] cavity and almost see the process of digestion." When Beaumont was transferred to Fort Niagara he took along his prize patient, who by then was ambulatory, and continued his studies.

Beaumont discovered that under the stimulation of entering food, certain juices oozed into the stomach and that the food disintegrated under the influence of these juices. He siphoned out some of the liquid and found that it was able to disintegrate food even outside St. Martin's stomach. There was no one with whom to share this exciting discovery. Beaumont continued his lonely studies patiently.

We must not visualize Beaumont's life as one of quiet ease or imagine that he could pursue his researches in a calm and serene atmosphere. He had his post duties, but worse still, his guinea pig began to realize his worth and became more and more demanding. He became tired of the diet which Beaumont fed him through his mouth and through the unnatural aperture and he began to supplement it with liberal quantities of whiskey.

Finally, three years after the accident, he ran away to his Canadian backwoods. Beaumont was brokenhearted at the abrupt end of his engrossing experiments, but four years later St. Martin, a newly acquired wife, and two little St. Martins, joined him again. In return for the support of the whole family Beaumont was permitted to resume his experiments.

He wanted to study the contents of these juices of the stomach, but, realizing that he was inadequately trained for the task, he tried to enlist the help of other physicians. He took St. Martin to New York but found, as he later wrote, that the doctors there "had too much personal, political, and commercial business on hand to turn their attention to physiological chemistry." He went to Yale, where he was advised to ship a bottle of St. Martin's stomach juices to the great chemist Berzelius, in Sweden. This, he did, but there is no record of what became of it. If it did arrive in Sweden, it must have been in an advanced state of putrefaction, fit only for the slop jar.

Beaumont continued his studies on the increasingly unmanageable St. Martin, alone. This untrained, lone army surgeon, without equipment, without encouragement, and at his own expense, carried on his experiments testing the influence of hunger, thirst, and taste on the secretion of digestive juices. He antedated, by many years, the Russian physiologist Pavlov, who later repeated many of these experiments on dogs with artificial stomach openings.

Beaumont's discovery of the potency of the stomach's juices in disintegrating food, even outside of the body, swept away all the previous notions on the mechanism of digestion. The stomach is indeed not a grinding mill nor a fermenting vat, but an organ which can make a magic solution to dissolve and disintegrate the entering food. Several years later the first component of the juice to be identified was called pepsin. It is an enzyme which splits proteins into their constituent amino acids.

How does pepsin perform its work? As we saw in the previous chapter, hundreds of amino acids are joined together to form a protein molecule. The couplings between amino acids are made by the shedding of water molecules, the amino acids being grafted together at the sites from which the water molecules are split. By boiling in strong acid the forces which keep the amino acids together are broken and, at the same time, molecules of water are inserted to patch up the shorn sites. Thus the original, intact amino acids are reconstituted. Pepsin achieves precisely the same thing; it, too, breaks bonds and adds water. How it manages to do this is the most intriguing, the most fundamental, and the most baffling problem facing the biochemist.

How very fundamental to life enzymes are can be realized if we consider the source of energy for all living things. The sun pours vast amounts of heat and light on our earth. The sun machine rotating in the optometrist's window, converts this energy by means of its black and silver vanes directly into motion. No living thing can do that. Living things use the sun's energy only through the mediation of enzymes. Only green plants are able, through their enzymes, to convert the sun's energy cascading on them into a different form of energy—chemical energy. The enzymes in the cells of all other living things can, in turn, use this stored energy for their daily needs. The following chemical reaction is the cornerstone on which all of life is built:

$$\text{CARBON DIOXIDE} + \text{WATER} + \text{ENERGY} \xrightarrow{\text{ENZYMES OF PLANTS}} \text{SUGAR} + \text{OXYGEN}$$

The 20 percent of oxygen in our atmosphere is testimony of the vast extent to which this reaction has been going on in the earth's history. All of the free oxygen in the atmosphere has accumulated from this reaction. No free oxygen could be here otherwise; it is so active in combining with other elements, that when the earth was a hot, molten, "ball of fire," none of the oxygen could have escaped combination with the other elements. The reverse of the above reaction is the battery, supplying the energy for the functioning of every cell of every living thing.

$$SUGAR + OXYGEN \xrightarrow[\text{IN CELLS}]{\text{ENZYMES}} CARBON DIOXIDE + WATER + ENERGY$$

If we throw a pound of sugar into a burning stove, it will burn, releasing exactly the amount of heat the sugar cane packed into it. The enzymes in our cells, however, perform this reaction in a controlled, slow process, releasing exactly the same amount of energy, gradually. (This happily slow reaction will be described in detail in Chapter 4.) We are lucky that the enzymes can do it in this manner, for otherwise, we would go up in a puff of smoke after a heavy meal.

Many biochemists have been attracted to the study of enzymes. Scores of different enzymes have been discovered. There are enzymes which break down proteins and others which break down fats; some enzymes have been shown to be essential for the sending of nerve impulses; the task of still other enzymes is the building up of body tissues. In every function of the body, a host of enzymes are involved.

We believe, that for every chemical process which takes place in a living cell—and there must be thousands of these—there is a separate system of enzymes. They are all remarkably specific. If there is the slightest change in the material on which the enzyme functions, the so-called substrate, the enzyme becomes impotent against it.

That enzymes can function outside of the cell was shown only about fifty years ago. This milestone is the monument to the Buchner brothers who demonstrated, in 1897, that sugar can be fermented to alcohol and carbon dioxide not only by yeast cells but also by water solutions of disintegrated cells, in the complete absence of living yeast. The name enzyme originated with this discovery. Enzyme is something *en zyme*—in yeast.

How purposeful and planned the achievement of the Buchners sounds. From reading this description of their work the reader probably visualizes the brothers working feverishly in their laboratory to prove an inspired hypothesis: that yeast juice will ferment sugar just as well as the living yeast does. Actually, the brothers, like many other experimental scientists, stumbled onto their discovery by sheer chance. They were trying out extracts of yeast as a medical concoction. Since they were going to feed it to patients, they could not use the usual poisonous preservative agents. So they turned to an old wives' remedy for the preservation of their extracts. It is well known to any one who makes jams or fruit preserves that a high concentration of sugar acts as a preservative. The Buchners added large amounts of sugar to their yeast extracts and the solutions began to ferment.

Actually, it was Beaumont who first saw enzymes working outside a living organism. Seventy years before the Buchners chanced upon their discovery he was digesting foods with St. Martin's cell-free stomach juice.

However, the meaning of great accidental discoveries, such as that of the Buchners, cannot be recognized until the time, or rather the scientist's mind, is ripe for it. A great deal of knowledge had been accumulated in those seventy years: the paralyzing concept of vitalism had been abandoned; Pasteur explored the nature of fermentation. The whole scientific atmosphere was favorable to the search for a chemical and mechanical interpretation of living processes. Only when steeped in such an atmosphere could the Buchners recognize the meaning of their accidental discovery.

Since the time of the Buchners, biochemists have extracted scores of different enzymes from a variety of different cells. Every such enzyme solution contains proteins. Slowly the suspicion grew that all enzymes *are* proteins. Before 1926 chemists were divided, however, on whether the proteins in the enzyme solutions were really the enzymes. There was a school of thought, particularly among German biochemists, maintaining that the proteins in the enzyme preparations were impurities, and the enzymes were elusive, smaller molecules, present in minute amounts. But in that year, James B. Sumner, at Cornell, was able to isolate an enzyme in a pure form, and it *was* a protein.

Sumner's achievement is so important that he was recently awarded the Nobel Prize for it. He had been studying the enzyme which breaks up urea into ammonia and carbon dioxide. This enzyme is called urease. The naming of enzymes is simple and uniform. To the name of the substance on which the enzyme works, is attached the suffix *-ase*. The scientist who first discovers the existence of an enzyme has the privilege of naming it.

Sumner chose a beautiful enzyme for his studies. The source, certain species of beans, is cheap—he grew the beans himself. The material on which the enzyme acts, urea, is also cheap. Furthermore, the enzyme produces ammonia and carbon dioxide, two of the easiest substances to assay. This in turn, makes the determination of the potency of the enzyme gratifyingly simple. The more ammonia a given weight of enzyme can produce from urea, the more potent it is.

Sumner isolated urease in a pure crystalline form. How does the biochemist go about such a task? He grinds up the source of his enzyme—in this case jack beans—with water, and obtains a thin brownish soup. His obvious question is, where is the enzyme? Is it in the soluble extract; or is it in the insoluble bean grinds? To decide, he adds urea to a small portion of each. The extract promptly begins to tear urea apart into ammonia and carbon dioxide; the insoluble bean debris is impotent. The enzyme is in the extract. But how much enzyme? A painstaking measurement of the ammonia that a certain volume of the extract can produce reveals the potency of the enzyme. The chemist now has his enzyme in solution but probably hundreds of other substances must be there along with it. So he begins the long, tedious task of concentrating the enzyme. He tries to throw out of solution, by means of various chemicals, either the enzyme or some of the contaminating substances. In every case both the substances thrown down and the solution remaining behind must be tested for enzyme activity. In every case the amount of ammonia formed from urea by each new preparation must be measured. As more and more inactive material is removed, the preparation becomes more and more concentrated—and smaller amounts of it will break down larger amounts of urea. After years of work and hundreds of treatments, Sumner obtained an enzyme preparation which was a crystalline protein. It is characteristic of organic substances that they do not crystallize until they are quite pure; the impurities intrude and prevent the formation of crystals. Obtaining a crystalline, pure protein which was a highly active enzyme was a great achievement. It established that at least one enzyme—urease, is a protein.

Since then, dozens of other enzymes, including pepsin, have been isolated in pure crystalline form, and every one of these, also, proved to be a protein.

It took 100 years to show that St. Martin's stomach juice owed its ability to split proteins to another protein, pepsin.

But the most spectacular development, stemming from Sumner's work, is the isolation in a pure crystalline form, of disease-producing viruses, which turn out to be protein molecules.

Now the biochemists know what enzymes are, and what they do. But *how* do they function? How does one protein molecule—the enzyme—pry apart the constituents of another protein molecule, the substrate? Of this we know next to nothing. We are attacking the problem from several fronts, but the more we learn about them, the more we realize how complex the problem and how far off is the solution. Research in the field of enzymes is not unlike climbing in a strange mountain range. From a distance a peak seems near. But, as the climber proceeds, he finds hidden gullies, gigantic rock piles, and rings of smaller ridges guarding the peak, and the more he climbs, the more territory he covers, the more distant and unattainable the original peak appears.

Let us look at some of the small ridges that have been conquered in recent years. A good deal of work has been directed toward the stopping of the activity of enzymes by so-called inhibitors. We try to learn how enzymes act by learning what agents stop their action. A very small amount of cyanide stops the action of several enzymes. All of the enzymes that are inhibited by cyanide contain considerable amounts of iron. It is well known to the inorganic chemist that iron and cyanide combine into a very tightly knit compound which leaves practically no free iron in solution. Cyanide inhibits these enzymes by siphoning off their iron. (That is why cyanide is such an effective poison.) Biochemists now know that iron is essential for the action of these particular enzymes. But are we any closer to our ultimate goal? Hardly.

The study of another type of inhibition of enzymes— competitive inhibition—has not brought us much closer to the object of our quest but has yielded a whole new battery of drugs to man in his fight against bacteria: the sulfa and other drugs. (The mountaineer may never reach his peak but he may find valuable mineral deposits.)

The work of Ehrlich on the development of salvarsan is well known but it merits continued attention because it is the fountain from which flowed the sulfa drugs, penicillin, streptomycin, and other aids to therapy. Ehrlich originated chemotherapy. To facilitate the recognition and classification of bacteria, he subjected them to various stains and dyes. They exhibited highly individual tendencies: some were stained by one dye and not another, or, more interesting still, in some cases only part of the bacterial cell was stained.

These erratically staining bacteria guided Ehrlich in his search for new drugs. A dye stains a cell by entering into a chemical union with its contents. Since there is such a profound demarcation, even within the same cell, between staining and nonstaining areas, it is entirely possible, argued Ehrlich, that there may be some poisonous chemicals which will selectively combine with microorganisms, damage them alone, and leave the tissue cells of the host unharmed.

He had spectacular success in exorcising the parasite which causes syphilis, following this principle. He and his associates kept making arsenic-containing organic molecules, almost at random, until they hit the bull's-eye with the celebrated "magic bullet," salvarsan. This compound kills the syphilis parasite by poisoning some of its enzyme systems. Fortunately, in the doses used in therapy, it is relatively nonpoisonous to humans.

Ehrlich's brilliant discovery nurtured the hope that new chemicals might be found which may be equally effective against other parasites which plague us and which are unaffected by salvarsan. The method of the search was the same as Ehrlich's—patient testing of each compound the researcher could lay his hands on. It was prospecting among organic compounds for new drugs instead of for gold in sand.

There were no guiding principles in the search. Each compound was tried on test-tube culture growths of various bacteria, and if any showed promise by killing the bacteria, they were tested on mice or other experimental animals. For many a drug is effective in the test tube, but useless in the whole animal, either because it is too poisonous or because it is rendered harmless to the bacteria by conditions in the animal. The work is slow and tedious. For example: a hundred mice might be injected with an identical dose of a virulent strain of streptococci—the little beasts which cause "strep" throats. In addition to this injection, fifty of the mice might receive the same dose of the drug which killed those streptococci in the test tube. In transferring from the test tube to the mouse the dosage of the drug is calculated by proportion. If one milligram of the drug killed the organisms in 1 cc. (about 1 gram) of broth, a mouse weighing 30 grams would get 30 milligrams of the drug. Then all that is left to do, after an impatient night, is to count the dead mice in each group. If the majority of the drug-protected mice survive, while the others perish, the drug is effective.

In 1935, twenty years after Ehrlich's death, Domagk, a German physician, struck gold in a dye called Prontosil. The new drug passed all the preliminary tests with flying colors: it protected mice against the streptococci, and it was harmless to the mice. The first human saved by the drug was Domagk's own young daughter, who had come down with a severe case of streptococcal blood poisoning. At that time, the physician could only lance the wound where the organisms made the breach, and hope that the body's natural defenses would rally and exterminate the invading cocci already in the blood stream. When his daughter continued to sink, Domagk fed her large doses of Prontosil. She rallied and recovered.

This melodramatic success was the first of a series of spectacular demonstrations of the value of the new drug. Those must have been hand-rubbing days at the I. G. Farben drug cartel, for Prontosil was patented, and the stage was set for the world monopoly of this potent drug.

"Fortunately for the world, however, Tréfouel and his colleagues in Paris soon showed that Prontosil acted by being broken up in the body with the liberation of sulfanilamide, and this simple drug, on which there were no patents, would do all that Prontosil could do." The quotation is from Sir Alexander Fleming, the discoverer of penicillin, the production of which the English scientists made available to all—in Sir Alexander's words—"without thought of patents or other restrictive measures."

Tréfouel's discovery—one of the few effective blows by a Frenchman against the Germans in that decade—gave tremendous impetus to the search for new antibacterial agents.

Sulfanilamide is a relatively simple compound; a competent sophomore in chemistry can make it. Furthermore, the organic chemist can make variants of it with ease. Onto the structure of sulfanilamide he hangs a variety of groups of atoms, and with high hopes he hands the new products over to the bacteriologist for testing.

Hundreds of altered sulfa drugs were made; some were better than the original sulfanilamide or Prontosil, others proved useless. But the search was still hit or miss. A guiding principle, an insight into the mechanism of the killing of the bacteria was lacking.

The English scientists Paul Fildes and D. D. Woods proposed an attractive theory. It is based on a theory of Ehrlich—the lock and key theory—which that remarkable genius had proposed for the explanation of the mode of action of enzymes. The history of the lock and key theory repeats the weary pattern of the reception of new ideas. Ehrlich was ridiculed by his contemporaries, but a new generation of scientists returned with admiration to the much abused theory, and used it eagerly.

This will continue to be the fate of new ideas as long as there are individuals who can forge ahead too many years in advance of their contemporaries. Not all new ideas or theories are sound; only time and work can weed out the barren from the fruitful in the garden of ideas. If only we could be more astute gardeners!

Ehrlich pondered the specificity of enzymes. Why does an enzyme act on one substance and not on another? He theorized than an enzyme and the substances it can alter must fit into each other as a key fits a lock, and that the possibility of such a union determines whether the enzyme can function on a substrate.

Fildes and Woods extended the theory to its next logical step. What would happen if another substance, which simulates in appearance the normal substrate, would fit into the lock of the enzyme molecule? The key may fit, but not completely; the enzyme mechanism may jam. They pointed to the possibility that sulfanilamide may simulate in structure some substance, X, in the diet of bacteria. Thus, sulfanilamide may crowd out the dietary essential X from the lock of the bacterial enzyme.

But what is X? It was already known that an extract of beef liver can protect bacteria against the sulfa drugs. If, to a suspension of bacteria, liver extract is added along with sulfanilamide, the drug is made impotent; the bacteria flourish.

A relentless hunt was started to track down the substance X in the liver. The suspension of finely hashed liver was subjected to a variety of chemical manipulations to determine, for example: Is X soluble in alcohol? Can X be thrown out of solution by adding chemicals which invariably throw down proteins? In every case the material before treatment and each fraction obtained from the chemical manipulation had to be tested for its ability to overcome the toxic effects of sulfanilamide. Casting this chemical dragnet is dull and tedious work but the hope of tracking down the active material spurs lagging spirits, and occasionally, the researcher's patience is rewarded. Since unsuccessful searches are seldom presented to the lay reader, solutions of problems of this kind must sound monotonously simple. They are far from it. Sometimes, years of exhaustive—and also exhausting—intellectual and physical labor yield nothing, and the search is sadly abandoned. However, this search was fruitful. The substance was isolated in pure form and to everyone's amazement it turned out to be a well-known chemical compound, para-aminobenzoic acid. This laboratory reagent is X, the essential substance, and it is, indeed, a vitamin for bacteria.[1]

The chemical structures of para-aminobenzoic acid, nicknamed PABA, and of sulfanilamide are strikingly similar. The conjecture which launched the search was beautifully confirmed. The enzymes of the bacteria may mistake sulfanilamide for PABA. The enzymes receive sulfanilamide, but since it is not completely the same as PABA there is soon confusion in the cell. Enzyme mechanisms stall, the bacteria cannot grow and cannot reproduce. If an extra dose of PABA is given to the bacteria at the same time as the sulfanilamide, the bacteria are no longer overpowered by the sulfanilamide, and continue to live. If the dose of sulfanilamide is again increased sufficiently, the bacteria once more will not grow. There is a definite numerical relationship between the amounts of PABA and sulfanilamide, which determines whether certain bacteria can live or not. Growth of a certain species of bacteria may be stopped if, in the fluid where they live, there are 1,000 molecules of

[1] Chemists should now treat every chemical on their shelves with renewed respect. Who knows what biologically potent materials may be hiding in those bottles?

sulfanilamide to one molecule of PABA. Their ability to grow is restored if the PABA concentration is increased to a ratio of 2:1,000. The various sulfa drugs differ in the PABA-sulfa ratio which will stop bacterial growth. For those bacteria which required 1,000 molecules of sulfanilamide to prevent growth, 10 molecules of sulfathiazole will suffice. Therefore, a patient invaded by these bacteria needs to be dosed with much smaller amounts of sulfathiazole than of sulfanilamide.

Biochemists visualize the sulfa drugs as competing with PABA for the favors of some enzyme in the bacterial cell. Such inhibitors of enzymes and of bacteria are called competitive inhibitors. Studies of competitive inhibitors yield just a tiny glimpse of the enzymes' work. We conclude that the enzymes must combine with a substance like PABA. The combination is probably chemical in nature. There must be a pre-existing "lock pattern" in the enzyme into which PABA must fit, and into which sulfanilamide also fits. However, in a later step in the process, sulfanilamide must jam the machinery.

The concept of competitive inhibitions also provides us with a rational approach to the search for new drugs. We need no longer try compounds at random. There is now a guiding principle in fashioning new "magic bullets." We try to make compounds similar in chemical structure to substances which are essential for the parasites, hoping that the new compound may be a competitive inhibitor in the parasite, without injuring the host. We have already found several new drugs with the aid of this new divining rod. But so far none of them competes with penicillin or some of the sulfa drugs.

Our new knowledge of competitive inhibition of enzymes

illuminated an old, old problem in nutrition. Pellagra, a disease which is prevalent among the underprivileged population of our South, was associated, for a long time, with eating corn instead of the nutritionally better wheat and rye. There is a biochemical adage: "The 'three m's': maize, molasses, and meat (mostly fat pork) will produce the 'three d's'—dementia, dermatitis, and diarrhea"—the symptoms of pellagra. Joseph Goldberger, of the United States Public Health Service, showed that a better diet of fresh lean meat, eggs, and milk, will cure pellagra. Later, it was shown that what is lacking in the "three m's" is one of the vitamin B's, niacin. Some time ago, D. W. Woolley, who is an outstanding American biochemist in the field of competitive inhibitors of enzymes, proved that corn not only is lacking in niacin, but, in addition, it actually contains something else which brings on pellagra! He kept mice on a diet, just adequate in niacin, to protect them from succumbing to the mouse version of pellagra. But if to this diet he added certain extracts made from corn, the mice did get pellagra. These extracts from corn contain something, probably a competitive inhibitor of the niacin, whose presence induces an apparent shortage of niacin in the animal. It seems that nature entered into a conspiracy against the corn eaters.

Why don't the animals such as fattening cattle, whose diet is largely corn, come down with pellagra? There is an answer to this question in the next chapter.

While our knowledge of the intimate mechanism of enzyme action is very limited, we do have much information about the different types of enzymes with which living things are equipped. Of all the enzymes, there are none more fascinating than those that produce light. Yes, "The fireflies o'er the meadow," which "In pulses come and go," owe to an enzyme the distinction of being lifted by the poet from the obscurity of thousands of other insects. They possess an enzyme which produces light! The firefly is not the only creature gifted with its astonishing flashlight. A variety of species, from the microorganisms that light up the wake of ships at sea, to large deep-sea fish which have lanterns to illuminate the otherwise eternal darkness of their hunting grounds, all use enzymes for the generation of light.

The enzyme has been named luciferase by E. Newton Harvey, the biologist to whom we owe much of our knowledge in this field. The light produced by luciferase is a cold light. The enzyme is able to generate light without wasting a good deal of energy as heat. Mechanically, we are no match for the firefly. Even the fluorescent lights, which are a great improvement over the hot-filament bulbs, generate some heat.

The most wonderful of these lantern-carrying creatures are some fish, which, though unable to make light themselves, carry around millions of light-producing microorganisms in little pockets under their eyes. When they want their lights dimmed, or put out, they cover the pockets with a convenient lid.

The purpose of these devices is not only illumination; the firefly has its light on its abdomen. Some deep-sea fish have light pockets along their sides, in characteristic porthole-like patterns, which vary from species to species. It is thought that members of the opposite sex of the same species recognize each other by these patterns.

Some animals use enzyme systems for attack or self-defense. The electric eel can discharge several hundred volts, over and over, to incapacitate its prey or enemy. The largest of these creatures, captured in the Amazon river, grow to five or six feet in length. About three fourths of the body is the electric organ which generates electricity with its remarkable enzyme system.

The electric organ of the eel is crammed with the same enzyme which is found in the nerve tissues of other animals, including man. Since the nerve impulses are relayed through electrical charges, the appearance of the same enzyme in these two entirely different tissues is not too surprising. It is one more illustration of what Emerson described as the poverty of nature. "And yet so poor is nature with all her craft, that from the beginning to the end of the universe, she has but one stuff . . . to serve up all her dreamlike variety. Compound it how she will—star, sand, fire, water, tree, man—it is still one stuff and betrays the same properties." (Exception must be taken to considering nature "poor with all her craft." Indeed, she is astoundingly rich in versatility, achieving, as she does, infinite variety with "one stuff.")

The venom of snakes is another example of assault by enzymes. There are two different types of snake venom. While both of them attack the blood of the prey, their mode of action is different; one causes the disintegration of red blood cells; the other clumps the red cells together. Since intact corpuscles are essential for the transportation of oxygen to outlying tissues the victim of snakebite will virtually suffocate for lack of oxygen.

Further explorations in enzymes will yield rich rewards. Scratch almost any problem in medicine and you expose a problem in enzymes: diabetes is a derangement of the enzymes which tackle the release of energy from sugars; cancer—a disease of unnaturally rapid and shapeless growth—is probably the result of some monstrous blunder in the functioning of enzymes; aging may yet be shown to be due to the slowing down or inhibition of some pivotal enzymes.

Whether a cut in the skin will become infected is decided, not only by the abundance of bacteria around the wound, but also by the abundance of antienzymes present in the blood. (These investigations are still in their infancy, so we do not yet find anti-infection enzymes on the shelves of the neighborhood drugstore.)

It was discovered that some bacteria produce a very useful enzyme which can dissolve the protective surface coatings of their prey. Using these enzymes as a battering ram, the bacteria can penetrate their prey all the more easily. The enzyme had originally been named hyaluronidase but this was mercifully changed to invasin. Invasin is inhibited or rendered ineffective by an antienzyme, anti-invasin, which is normally present in the blood stream but which is deficient in patients suffering from severe bacterial infections. Too little information is available to tell how these antienzyme shock troops of the body repel the invading enzyme; whether they are enzyme inhibitors or enzymes which gobble up other enzymes. Time and more research will tell.

At the time of the original discovery of invasin, the enzyme appeared to be of only academic interest. But research sometimes takes unexpected turns. Goethe's dictum "what is true is fruitful" has been often borne out in the growth of science. An apparently modest little discovery may swell through the work of the original discoverer, or of others, to impressive proportions, providing excitement and joy to all researchers—particularly to the one who made the initial observation. What may develop from studies of invasin and anti-invasin in the next decade is impossible to predict now. A totally unforeseen practical benefit has already accrued from it: the remedy of a certain type of sterility in humans.

To make the appointed task of the sperm easier, the human sperm fluid is richly stocked with the very same enzyme the bacteria produce, invasin. In some cases of sterility it was found that the male is deficient in the production of invasin, and the sperm, unaided by this mighty trumpet, is impotent before its Jericho. The shortcoming has been effectively overcome in a number of cases by the appropriate use of preparations of the enzyme from bull testes.

So far, discussion has been restricted to what has already been accomplished in enzyme chemistry. But let us now take a peek into the enzyme chemistry of the "Brave New World" of the future. The writer has a pet enzyme inhibitor, alas, yet to be discovered, which may solve a social problem that has distressed men of good will and of ill will for generations.

The extent of the pigmentation of the body is determined by enzymes. As one of our amino acids, tyrosine, is combined with oxygen, it forms dark-colored products, the so-called melanins, which produce the color of the hair, eyes, and skin. We see the result of the complete absence of this enzyme, through an error of heredity, in albinos with snow-white hair, fair skin that cannot tan, and unpigmented pink eyes. On the other hand, we have the more abundant pigmentation of the colored races. According to our best authorities, "No pigments other than those found in the whites are encountered in the dark races," and therefore, "the colored races owe their characteristic color only to variations in the amount of melanin present"—and, in turn, to an overactivity of the enzymes which produce melanin. The difference between the blondest and the darkest of humans is only enzyme deep.

It is entirely within the realm of biochemical possibility that someday a specific inhibitor will be found which, when fed, will slow down the enzymes which produce the melanin pigments, enabling us to lighten skin pigmentation at will. Or, on the other hand, we might find an enzyme *accelerator*, as well, which will enable us to darken lighter skins. For, after all, who is to decide what is preferable? The judicious use of an enzyme inhibitor and accelerator may thus someday achieve a Utopia of color.

3 Vitamins

The enzymes' helpers

WHY do we need vitamins? Why does the absence of a minute dust of white powder from a sailor's diet cripple him with scurvy? What can the vitamins do in our cells to make us their slaves?

The question of the role of vitamins in the cell will be answered against the background of an acrimonious scientific controversy which raged for almost eighty years before it was completely resolved.

Solution to a problem in science does not pop out of the head of one genius, like Minerva out of Zeus's forehead, fully developed and completely integrated into the rest of the body of scientific knowledge. Our understanding of the role of just one vitamin required the intellect and labor of four generations of scientists!

Before we get on to the controversy let us introduce the two protagonists who started the feud. In one corner is the champion, Louis Pasteur, the greatest biological scientist of the nineteenth, perhaps of all, centuries. At the age of thirty-eight, when this controversy started, he was already a famous chemist with a list of brilliant achievements and high honors to his name. About this time he was leaving the field of the chemistry of crystals and was laying the foundations of bacteriology as a science. Later he studied the "diseases of wine" under municipal sponsorship and the diseases of silkworms under commission of the French ministry of agriculture.

From his studies, Pasteur concluded that the "diseases of wine" were produced by bacteria which contaminate the juice of the grape, and whose nefarious by-products were assaulting the French palate. He prescribed appropriate remedies: warming the unfermented juice to destroy the trespassing, undesirable organisms. Had he gone no further, Pasteur would have been universally hailed by his countrymen as a savior of his country second only to Joan of Arc. (No doubt there must have been many who would have rated delivery from the despoilers of wine even higher than delivery from the British.) But Pasteur did go further. He widened his researches and his conclusions and announced that some human diseases, too, are produced by bacteria. With this statement he came into a head-on collision with some members of the French Academy of Medicine. There ensued a series of celebrated polemics— including at least one invitation to duel—in which Pasteur showed himself not only a man of genius but also a man of iron will, ready to fight for truth as he saw it through his microscope. We get a glint of the steel in the man, in this earlier, less publicized controversy.

The fermentation of sugar into alcohol and carbon dioxide had been known for a long time, but the motivating force which induces fermentation was unknown.

The German chemist Liebig had proposed the most popular theory of the time. Fermentation was supposed to be produced by the last vitalistic "vibrations" of dead biological material. Decades had passed since Wöhler's synthesis of urea, but vitalism still dominated the minds of many scientists.

Pasteur concluded from his own studies that, on the contrary, fermentation is the normal function of living yeast cells and that it proceeds apace with the growth of yeast cells. As irrefutable evidence for his theory, Pasteur published, in 1860, a paper entitled "A Note on Alcoholic Fermentation."

In this classic he showed that from a solution of mineral salts, ammonium salt, sugar, and a very small seeding of yeasts, "the size of a pinhead," he obtained both fermentation—he zestfully described the copious evolution of carbon dioxide—and a lush growth of healthy yeast cells.

This was electrifying news. Here was evidence that a living thing, a yeast cell, can grow and reproduce, in a medium completely devoid of any mysterious vitalistic substance. Only sugar, minerals, and ammonia were needed. Such a fundamental experiment was bound to be repeated.

And now, the challenger: Justus Freiherr von Liebig was the dean of German chemists who, in 1869, at the age of sixty-six, could look back on a life rich in achievement in organic and in agricultural chemistry. Liebig announced that he could not repeat Pasteur's experiment! Furthermore, he literally insulted Pasteur by suggesting that he deceived himself by mistaking for yeast some stray molds growing in his flask. Pasteur, the greatest living expert in bacteriology and a handy man with a microscope, unable to distinguish a yeast cell from a mold filament! He replied with characteristic pungence: "I will prepare, in a mineral medium, as much yeast as Mr. Liebig can reasonably ask, provided that he pays the cost of the experiment." Furthermore, Pasteur invited Liebig to come to his laboratory so that he might repeat the experiment in Liebig's presence. Liebig was, for those days, an old man and he died four years later, in 1873, at the age of seventy, without accepting Pasteur's defiant challenge.

The first round went to Pasteur on points.

The next important development in the controversy came in 1901, six years after Pasteur's death. Wildiers, at the University of Louvain, calmly restudied the problem of raising yeast cells in a mineral medium. He found that the crux of the problem was the size of the droplet of yeast cells used to inoculate the sterile "mineral" broth. Pasteur said he used a droplet the size of a pinhead. Unfortunately that was not a very exact prescription. Just as medieval philosophers are said to have debated on how many angels could stand on the point of a pin, bacteriologists began to debate the *size* of a pinhead.

Wildiers found that if the size of the inoculating droplet is made very small, yeast cells do not grow in Pasteur's medium. The few that grew were sickly looking, malformed little creatures. He reluctantly took Liebig's side in the controversy. But he went further than that. He showed that with a large droplet he could repeat Pasteur's successful experiments. Furthermore he could use a very small inoculum of live yeast cells plus a large droplet of sterilized yeast cells, in which all the organisms were killed, and, adding these two to the mineral broth, he obtained lush yeast growth. He concluded that there is something other than live cells in the large inoculum which the yeast must have for growth. He called this something bios—from the Greek word for life. Wildiers found that bios is present in a variety of substances. A sterile extract made from meat or from egg yolk, added to Pasteur's broth, enabled the yeast to grow from minute inocula.

There was something present in these extracts which was indispensable to the growth of yeast cells. Thus Wildiers demonstrated the existence of vitamins, long before the term was coined. He tried to isolate the bios. But the current techniques of chemistry were not up to the task.

One might expect that progress in the bios problem should have been rapid after this. Far from it. The controversy really began to rage in earnest. At first it was denied that there is such a thing as bios. Experiments were brought forth showing that yeasts do not need bios. These may very well have been sound experiments, for we know today that different strains of yeasts do have different nutritional requirements. Some antibios crusaders held that bios was not essential; it merely overcame the poisoning of yeast by copper. Verifying Pasteur's famous dictum—"Nothing is so subtle as the argumentation of a dying theory"—some went even so far as to say that yeast grew better with extracts of meat, not because of bios, but because of some other substance. These bacteriologists, it has been said, were "qualified to join the Last Ditch Bacon Club, which holds that Shakespeare's plays were written, not by Shakespeare, but by some other, bearing the same name."

The bios contenders continued to wrangle; anyone who could think of nothing more productive to do could always show that some exotic substance did or did not contain bios.

This round belonged to Liebig, the challenger.

Before we go into the final round in the controversy we

must turn our attention to another area in science where big strides, which eventually led to the resolution of the bios problem, were being quietly made.

The knowledge that foods can remedy some human diseases is as old as recorded history. In some Egyptian papyri we find descriptions of the ritual by means of which the priests restored the eyesight of travelers, returned from prolonged trips in the desert. With appropriate incantations they fed to the afflicted the liver of a donkey sacrificed under suitable omens in the sky. In the Apocrypha there is the story of Tobit (Tobias, in the Vulgate), who lost his eyesight. His son Tobias was instructed to catch a monster from the sea and "anoint" his father with its liver. It is difficult to determine, at this distance, how the son might have interpreted the original term for "anoint." If he fed the "monster" liver to his father he practiced perfectly sound vitamin therapy. The lack of vitamin A in the human diet causes, at first, night blindness and later, almost complete blindness. The richest source of this vitamin is the liver of animals, especially fish.

The cod-liver oil industry ought, perhaps, to make young Tobias its patron saint, for he was the first to practice the trade. The use of fish-liver oils in therapy, in modern times was first mentioned in 1782, when the English physician Robert Darbey wrote, "an accidental circumstance discovered to us a remedy, which has been used with great success . . . but is very little known, in any country, except Lancashire. It is the cod, or ling liver oil."

How very recent is our knowledge of vitamins can be appreciated from the following quotation from a leading textbook on diet. "The chief principles in food are: Proteids

[Archaic name for proteins], Carbohydrates [sugars], Fats, Salts, Water." Not an inkling of anything else. That book was published in 1905.

Since the story of the discovery of vitamins has been often told it needs to be summarized only briefly. A Japanese admiral, Takaki, had a hunch that the beriberi with which sailors on long voyages were plagued, might be due to their poor diet at sea. He was a born experimentalist for, in 1882, he sent out a ship well stocked with meat, barley, and fruits, and indeed, no beriberi occurred among the crew. This was clear-cut evidence for the relation between diet and disease, but of course there was still no glimpse of what was lacking in the diet.

Fifteen years after Takaki's cruise a great stride was made by a physician of a Dutch penal colony in Java, Dr. Christian Eijkman. He noticed that hens feeding exclusively on polished rice—the staple diet of the natives— came down with a strange ailment. They were overcome by lassitude which progressed to complete paralysis followed soon by death. Eijkman was able to revive the moribund birds by feeding them the polishings from the rice.

This was a profound discovery. In the first place, here was an unequivocal demonstration that withholding a part of a food from the diet can induce a disease and restoring the same food can cure that disease. Furthermore, having an experimental animal in which a disease can be induced at will is always a great asset. (Our inability to induce pernicious anemia in experimental animals had been a tight brake on our progress against this disease.) When we have convenient, susceptible experimental animals, forays can be made against a disease from many sides: the internal changes at various stages of the disease can be studied; a variety of possibly dangerous medications can be tried out; once a medication is won, it can be standardized.

Eijkman realized that there is a substance in the outer coats of rice which is essential for health. He thought that in the absence of this material the utilization of sugar may follow an abnormal path in the body. Thirty more years of work were needed before Eijkman's prophetic insight in 1906, into the role of a "vitamine," was confirmed. That was the name coined for the substance in the rice polishings by Casimir Funk, one of the outstanding early workers in this field.

The result of Eijkman's work was the concept that in addition to an animal's diet of proteins, fats, and sugars, vitamins, too, are needed to secure healthy cells. New vitamins, a whole alphabet of them, were discovered, but their specific task in the cell remained unknown.

At the same time that Eijkman announced the results of his work, an important parallel discovery was made in England. The intimate relation between the two discoveries did not become apparent for thirty years. The new discovery was destined eventually to throw some light on the role of the vitamins and of bios.

The Buchner brothers were able to ferment sugar to alcohol and carbon dioxide, with cell-free enzyme extracts of yeast. (That was in 1897, almost forty years after Pasteur established the nature of fermentation.) Then in 1906, two Englishmen, Sir Arthur Harden and W. J. Young, performed a challenging experiment. They took an active yeast-enzyme solution, and passed it through a gelatin filter which was known to hold back very large molecules but to allow smaller molecules to go through. Neither the fraction that remained behind on the filter nor the fraction that passed through was able to ferment sugar! But if the two fractions were pooled, the solution was as good as before in fermenting sugar solutions. For successful enzyme action, then, two factors are needed: large molecules—and we now know that these are proteins—and some smaller molecules to assist the enzymes. These were named coenzymes. But what are coenzymes? Chemists began the tedious task of concentrating solutions of coenzymes with the hope of eventually isolating them. At the same time, other chemists were working on the isolation of vitamins, for example, vitamin B_1 from rice polishings.

Before the result of these parallel searches is announced, tribute should be paid to a patient Austrian chemist who helped tremendously every other chemist who has to work with minute amounts of substances.

The year was 1910. Fritz Pregl, a professor of chemistry in Graz, was investigating the constituents of bile. Patiently he isolated a couple of hundred milligrams (28,000 milligrams make an ounce) of pure crystalline material of unknown composition. The first step he had to take to establish the composition of his precious substance was to determine the amount of carbon and hydrogen it contained. The available methods for this task required the burning of from 300 to 500 milligrams of material.

The destruction of 300 to 500 milligrams of precious substance which took years of labor to accumulate would provide but one bit of information. Nothing would be left for the dozens of other determinations and manipulations that had to be performed before the complete structure of an unknown substance could be pieced together. Pregl rebelled. He spent the rest of his life perfecting and refining methods so that determinations could be done on one or two milligrams of material. He was spectacularly successful. After the First World War, chemists came to him from all over the world to learn his methods. These disciples, in turn, spread the gospel of microchemistry, for that was the name given to this new technique. Pregl was awarded a well-deserved Nobel Prize for his work. Biochemists were now equipped with sufficiently refined tools to go digging for substances which, like vitamins, were present in their natural sources in minute amounts.[1]

We are now ready to return to the last round in the bios controversy. In 1919 a young American biochemist Roger J. Williams published his doctorate, in which he stated that yeasts need "growth promoting substances" in addition to sugar, salt, and ammonia and these substances were identical with "the substances which in animal nutrition prevent beriberi." Yeasts and beasts need the same vitamin!

At the same time Williams's older brother Robert R. Williams, a chemist at Bell Telephone Laboratories, was struggling alone, at his own expense, with the isolation of this very vitamin B_1 from rice polishings.

The next dramatic development in the history of bios was the announcement in 1936, by the German chemists F. Kögl and B. Tönnis, of the isolation of bios in pure crystalline form. They obtained, after working up 500 pounds of dried egg yolk imported from China, 1.1 milligrams of a pure substance which they named biotin, in honor of the bios problem. Their method of isolation was the usual painstaking physical and intellectual labor, stretching over several years: subjecting the dried yolk, a rich source of

---

[1] Still further refinements of Pregl's techniques enabled our chemists on the Atomic Bomb Projects to master the chemistry of plutonium, from a few *thousandths of a milligram* of it. On the basis of that knowledge the vast plants for its large-scale production were built.

bios, to a variety of chemical separations; testing each fraction for its yeast-growth-promoting potency; discarding the inactive, concentrating the active fractions more and more, until the material was sufficiently purified to reward their labors by crystallizing in pure form.

But what is biotin and what is its function? Its actual chemical architecture was not decoded until six years later but its role in the cell was soon established by inference.

As a result mostly of the efforts of Robert Williams, vitamin B_1 was isolated in pure form and made available for study. R. A. Peters, in England, showed in 1936, that vitamin B_1 is a cornerstone in the structure of the coenzyme which assists one of the enzymes responsible for burning sugar in the cell. We should recall with awe Eijkman's prediction thirty years earlier of a similar role for this vitamin.

Now the specific chemical role of at least one of the vitamins was known. One of the products in the degradation of sugar in the cell is pyruvic acid. This acid was known to accumulate in the blood of patients suffering from beriberi. It was also known that there is an enzyme in most cells which causes the breakdown of pyruvic acid to carbon dioxide and acetaldehyde.

The enzyme which extracts carbon dioxide from pyruvic acid, needs a spark plug for its smooth functioning. The spark plug is a coenzyme made from vitamin B_1. If the animal does not receive sufficient amounts of vitamin B_1 its cells and body fluids become overloaded with pyruvic acid causing the external symptoms of beriberi, such as paralysis.

Between 1905 and 1936 biochemists whose main interests were coenzymes were busily gathering information. It was found that the coenzymes of fermentation by yeast were present, not only in yeast, but also in such varied materials as milk, animal organs, and blood. Whether the coenzyme preparation was made from extracts of yeast or extracts of frog muscle, it appeared to be identical. Yeasts and beasts need the same vitamins; yeasts and beasts have the same coenzymes.

These developments provide additional support for Darwin's theory of evolutionary ascent from some common origin. For here is biochemical evidence for the most intimate similarity between yeast and frog; they need the same vitamins; they make the same enzymes; they need the same coenzymes. It must not be inferred that frogs have evolved from present-day yeasts. The implication is that they both evolved from some common ancestral cell in which these basic enzymes and coenzymes were already present.

The coenzyme of yeast fermentation was isolated in 1935. It contains niacin which, just about then, was also shown to be a member of the vitamin B group. Thus another vitamin, whose absence in the diet of animals induces pellagra, turned up as a coenzyme. Man needs vitamins as assistants to his large variety of enzymes. And bios or biotin is needed by the yeasts, also, as a coenzyme.

The biochemist has so far extracted twelve different components from what originally was thought to be a single substance, vitamin B. These twelve B vitamins are grouped together for historical reasons only. In chemical make-up they have nothing in common. They vary in the elements of which they are composed and in the alignment of those elements; some of them have very simple structures, others are quite complex; their only similarity is their solubility in water. In function, too, they vary as much as in structure. While they all act as coenzymes, the chemical reactions at which they assist encompass the whole spectrum of biochemical processes.

The history of biotin offers an example of the elucidation of several apparently unrelated problems, once the chemist has a pure substance. Before biotin was isolated, a number of researchers reported the existence of several unknown factors that behaved like vitamins. The most interesting of these is the one called vitamin H. If rats are fed raw, uncooked, egg white as the source of their protein they do not thrive. At first a skin rash appears; then they lose their hair; then they become paralyzed and, if the diet is kept up, they die. They can be saved by cooking the raw egg white, by replacing it with another protein, or by feeding to them along with the egg white either egg yolk or beef liver.

The circumstances pointed to a deficiency disease, and a search was started for vitamin H. (The earlier letters of the alphabet had already been pre-empted.) At about the same time it was found that a certain microorganism whose habitat is the root of legumes needs an unknown substance in its diet or it perishes. The substance was named coenzyme R and the search for the substance to fit the name was started. There was still another observation: certain diphtheria bacilli need a well-known compound called pimelic acid for *their* growth. After biotin was isolated and made available, it was found that vitamin H was biotin, coenzyme R was biotin, and the diphtheria bacilli used pimelic acid for the synthesis of homemade biotin.

The disease caused by raw egg white turned out to be due to the deficiency of biotin. There is a substance called avidin [2] in the raw egg white which seizes the biotin and forms with it a tightly knit combination. The biotin cannot then be absorbed from the intestine. (Avidin loses this property when it is cooked.) The disease—egg-white injury—can be produced in humans, too. Volunteers who ate a diet in which raw egg white provided 30 percent of the calories developed the characteristic skin disease in two to three weeks. "This symptom disappeared, but in the fifth week one of the group developed a mild depression which progressed to an extreme lassitude and hallucination. Two others became slightly panicky. The only striking observation in the seventh and eighth week was a marked pallor of the skin." In the ninth and tenth weeks the skin rash reappeared. The subsequent symptoms are not known for the experiment was halted and the subjects revived by adequate doses of biotin.

The depression and hallucination of one of the volunteers is very significant. This is not an isolated case of the appearance of such symptoms as a result of vitamin deficiency. Volunteers existing on diets deficient in vitamin B_1 showed similar symptoms, and the dementia of pellagra is well known. (A fuller discussion of this will be found in the chapter on "The Brain: Cells that Think.")

Biotin, a vitamin which is essential to the health of yeasts and was discovered through research on yeasts, is apparently essential for the health of humans as well.

Biotin is effective in astoundingly low concentrations. It is one of the most potent of biological substances known. A rat needs only .03 micrograms a day. One teaspoonful of the crystals [3] would be enough to supply the daily needs

---

[2] A contraction of avid albumin.
[3] About 3 grams or 3,000 milligrams or 3,000,000 micrograms.

of 1,000,000 rats for 100 days. Since generally in the case of drugs there is a rough relationship between the dosage of the drug and the total weight of the recipient, an approximation for humans can be made, too. A man weighing 150 pounds is about 500 times as heavy as a rat. Therefore, the approximate daily need of biotin for a man is 500 times .03, or 15 micrograms. The teaspoonful of biotin would suffice for 2,000 men for 100 days.

A very interesting suggestion has been made concerning the relation between biotin and cancer. The hypothesis is interesting in itself but what makes it even more so is that it was made by W. L. Laurence, the science writer of the New York *Times*, who, after writing for years about the doings of other scientists, caught the spirit and made news for himself. There were early reports that embryonic tissue and tumor tissue, both of which grow very rapidly, are rich in biotin. Perhaps biotin is the cause of rapid growth. This may sound like very tenuous reasoning, but even flimsier hunches have, in the past, proved fruitful. Furthermore, in a field where all is darkness, even the tiniest gleam of light should be tracked down. Laurence offered some previously known phenomena of cancer which tied in with his hypothesis. Nothing is as welcome to the scientist as the finding of previously described data which he can marshal as evidence for his hypothesis and thus avoid the tedium of prolonged experiments. It must have been especially welcome to an eager scientist who could not perform an experiment.

Laurence cited 300 authenticated cases of spontaneous regression of cancer, where, for no apparent reason definite cancerous growths disappeared. Of these regressions, 100 occurred in patients who, while they had cancer, caught

some acute infection of microorganisms as well. Since the
need for biotin is widespread among microorganisms, Laurence suggested that perhaps the regression of the cancer
was due to the lapping up of all the biotin in the patients
by the infecting organism.

The next logical step was to study the effect of biotin
starvation on experimentally produced tumors. A group of
mice were kept on a diet of raw egg white. When they began to show signs of biotin deficiency, cancer transplants
were placed in them, surgically. At the same time normal
mice were subjected to similar cancerous transplants. If
biotin were a controlling factor the cancers should not
thrive in the biotin-starved mice. Alas, in both sets of mice,
the biotin-free and the biotin-rich, the tumors grew at the
same rate.

Another approach to the same problem is the use of competitive inhibitors of biotin to crowd out the vitamin from
the cancerous tissue, but in view of the failure just cited
this new attack does not hold much promise.

But let us return to a final survey of the bios problem.
We can now view it from a vantage point: the top of
the mountain of knowledge accumulated during the pre-
ceding fifteen years. Biotin is not bios. Something else
is the bios that Wildiers described. In the past fifteen years,
since biotin was isolated, great strides have been made
both in the chemistry of the B vitamins and in the study
of the dietary requirement of yeast. There are no less than
twelve different members of the vitamin B family. Yeasts
need five of these. The other seven they either do not need
or they can make themselves. The five needed by the yeasts
are the following: thiamin (B_1), inositol, biotin, pyridoxine

(B_6), and pantothenic acid. Of these the last one, discovered by Roger Williams, fits the description of the original bios best. For example, the bios described by Wildiers had to be a very sturdy compound to withstand the prolonged heating customarily used in those days for sterilization. Biotin would surely have been destroyed, but not pantothenic acid. Of course, all of this is merely of historical interest and as in the affairs of men, so in the affairs of yeasts, historical problems are not easily resolved. We do not know with certainty what strains of yeasts Wildiers used, or whether they were pure, homogeneous strains. There are hundreds of different yeast strains, and their dietary needs vary from strain to strain. Furthermore, we also know today that their dietary requirement varies with the time they are allowed to incubate. Given enough time they can make some of these vitamins themselves, among them biotin, but not pantothenic acid. But what name has been given to which vitamin is, after all, of no consequence. What matters is that an old problem has been solved, and during the course of its solution we acquired a great deal of new knowledge. And knowledge is our most valuable possession. Knowledge not only is rewarding in itself but also it often leads to new developments of practical importance. Fifty years ago the "bios problem" must have appeared ludicrous. Who but "impractical" professors would care whether yeasts need bios in their diet? But from the exploration of the dietary needs of yeasts and of other microorganisms, we learned of the existence of six new members of the vitamin B family. These vitamins improve our own nutrition, one of them is a potent drug against pernicious anemia, and finally, the ultimate in justification, millions of dollars have been made on them. The professors, alas, were not included in the last activity.

Since this book attempts to be a chronicle of ideas, not a purveyor of prescriptions, it will avoid admonition and advice on the choice of appropriate foods for obtaining the full quota of vitamins. Furthermore, we are deluged by information about vitamins these days. To cite a few of these fragments would be pointless; to do a thorough job is impossible. A recent book, *Vitamins in Clinical Practice*, contains 1,000 large pages. Moreover, there is not enough space here to relate the history of the other vitamins, although they are equally as interesting as that of biotin.

There are some aspects of our recent knowledge about vitamins, however, which have not yet been incorporated into radio commercials and which throw a light on some of the questions we have been asking.

In the previous chapter the question arose why cattle feeding on corn do not develop pellagra. Also, what is the reason for the large individual differences in vitamin requirements which are known to exist in different persons? The answer to both of these questions is the same: the production of vitamins by the microorganisms in the alimentary tract. The alimentary tract teems with microorganisms, most of them harmless, nonpathogenic. Not only are they harmless but, indeed, they are absolutely essential for the life of the cattle on their natural diet. The alimentary canal of the newborn calf (or child) is completely sterile. If the young calf's stomach and gut were to remain forever sterile the animal would have to be restricted to a constant diet of milk or it would perish. However, with its very first meal it acquires the founding fathers of an

enormous colony of what is euphemistically called its intestinal flora. The cow, even though it has a multiple stomach, does not have the ability to digest the cellulose which
makes up the largest part of its diet. The cow lacks the
enzymes to split cellulose. It would starve to death with
a stomachful of grass if it were not for its alimentary flora.
For these microorganisms are able to convert the cellulose
to smaller molecules. While doing this they keep for themselves some of the food and some of the sun energy that
had been packed into the cellulose. But they "live and let
live," and more than enough is left for the cow. This is a
marvelous cooperative enterprise. The cow gathers the
food and provides warmth for the little creatures. They
pay rent with their labor, for they not only help with the
cow's digestion but also provide much of its vitamin requirement. Many microorganisms can synthesize most of
the vitamins for their personal needs. As they die, their
cell's contents ooze out and the vitamins are absorbed by
the cow.

Man, too, plays host to huge colonies of microorganisms
in his alimentary canal. Many of these tiny creatures re-
pay this kindness with their homemade vitamins. That this
source can contribute a considerable portion of a man's
vitamin requirement was made evident by the production
of vitamin deficiencies in patients as a result of the prolonged feeding of sulfa drugs. The drugs not only killed
the bacteria in the patients' tissues but also wiped out their
vitamin factories by the indiscriminate destruction of the
alimentary flora.

The amount of vitamin an individual must receive from
his diet is dictated by two factors: his body's total requirement and the amount of vitamins his intestinal flora will

make for him. Microorganisms differ tremendously in their ability to make their own vitamins; some can make almost all of them, others can make none. Our knowledge of this relationship between man and the microorganisms he harbors is still very limited, but it will certainly explain the hitherto puzzling individual variation in vitamin requirement.

There are experiments now in progress to raise rats in an absolutely sterile environment so that they cannot acquire an intestinal flora. Hercules had no more difficult task getting the Augean Stables superficially cleaned, than have the scientists in keeping one cageful of rats free of microorganisms. From birth these rats must be kept in an absolutely sterile cubicle with the incoming air completely sterilized. Their water and their food must be sterilized, and all of these tedious tasks must be carried on, relentlessly, for months. A momentary breakdown in the air-circulating and sterilizing machinery might introduce a few bacteria to ruin months of work. After years of unsuccessful attempts, this project is just now succeeding.

There is an interesting sidelight on the possible role of intestinal flora in lengthening man's life span. At the end of the last century Metchnikoff, the Russian physiologist, was impressed by the large number of hale centenarians he found among inhabitants of the Balkan mountains. He found that sour milk was a staple in the diet of these folk, some of whom were doubling the Biblical three-score-and-ten life span. Their milk was soured by certain bacilli which convert milk sugar into an acid, lactic acid. Owing to Pasteur's influence, all scientists were very much preoccupied with the role of bacteria in health and disease. Metchnikoff conjectured that the large colonies of these lactic-acid bacteria in the intestinal flora may crowd out pathogenic organisms and thus promote longevity. The drinking of soured milk became a widespread fad. (Indeed, the bottles of Yogurt sold in Paris still carry the legend that its consumption is recommended by Professor Metchnikoff.) It would be nice to report now, with our newer knowledge of vitamin synthesis by intestinal flora, that there may be sound basis for enriching our intestines with lactic-acid bacteria and that the vitamins from the lactic-acid bacteria prolong the life of the Bulgarian mountaineers. Unfortunately, there is no basis for this. On the contrary, the lactic-acid-producing bacteria are among the least versatile of microorganisms in this respect. Unless their diet includes most of the vitamins these bacteria die.

With our present ignorance of the causes of aging the only prescription we have for longevity is what can be gathered from statistical studies of people who are blessed with it. Eat well, but not too much; relax; avoid infectious diseases; and above all, choose long-lived ancestors, for heredity seems to be the most important factor.

Vitamins are not cure-all elixirs. A living cell has awesome complexity; one minute component of it cannot be the be-all and end-all of its smooth functioning. On the other hand we have been prodigally destructive of our foods. In this respect, most other animals have been far more intelligent than we. There are indications that most animals choose vitamin-rich foods fastidiously; man, however, polishes his rice and bleaches his flour,[4] wasting or

---

[4] A certain type of bleaching produces such deterioration in flour that it induces convulsions in dogs when they are kept on an exclusive diet of bread made from such flour. Since bleached bread is not the sole protein in the human diet no harmful effects have ever been noted in humans. Nevertheless, that particular type of bleaching has been discontinued.

destroying his vitamins and other factors. He overpurifies his foods, for example sugar, until they are completely free of the minute amounts of vitamins or minerals originally present. Nor is he too quick to apply what the scientists teach him. Robert R. Williams reports that: "A bill to put a tax on white (polished) rice in order to discourage its use was first introduced into the Philippine Legislature in 1911. Although it has since been reintroduced several times, it has never succeeded of passage. As a result, incapacitation from beriberi is substantially as prevalent as it was thirty years ago, though we have known all that time how it could be prevented. The same condition prevails generally through the rice-eating areas of the Orient." Before we judge the ignorance of the Philippine legislators too harshly, let us consider what chance a similar bill on chemically bleached flour would have in our own Congress.

Nutritional Utopia will not be reached by legislation; its only chance is by research and education. Nor is the replenishing of the known destroyed vitamins the complete answer. We must not repeat the error of the prominent physicist who, at the start of the century, categorically stated that physics as a science is completed. Our knowledge of nutrition is far, far from complete.

How limited is our knowledge of all of man's nutritional requirement is indicated by the shortcomings of the army's concentrated rations. Our best knowledge went into the compounding of these concentrated emergency rations, the K, the C, and the "10 in 1." One of our leading nutritional experts, Carl Elvehjem of Wisconsin, kept monkeys on these rations. "All the monkeys," he reports, "that we have placed on the Army rations have failed rather rapidly. Some have been a little better than others, but I don't believe that you could keep any monkey alive for more than six months on any of these rations." Such rations were, of course, intended to be eaten for only short periods of emergency. We still cannot duplicate completely in a concentrated form a good, natural diet for prolonged feeding.

During the Second World War the diet of our English allies was strongly fortified with vitamins. But this was in addition to a wisely planned general diet. The diet of the average Englishmen of the poorer classes during the war, with all its shortages, was better nutritionally than before the war.

On the other hand, we shipped tons of vitamin B_1 to China too. A Chinese biochemist friend considered this the worst possible blunder. Vitamin B_1 is an excellent stimulant of the appetite. The poor Chinese, with his traditional bowl of rice a day, does not get much in the way of vitamins or food. But as the result of years of vitamin deprivation, his appetite, too, is limited. To stimulate his appetite, without providing for satisfying it with increased food rations and increased allotments of the other vitamins was hardly a solution of his problem.

And, finally, who shall be declared the winner in the Pasteur–Liebig controversy? This frankly prejudiced referee votes for Pasteur. Remember, Pasteur's whole thesis was that fermentation is the result of the living activities of yeasts. In this he was utterly correct. Further, he stated that yeasts can be grown in a mineral medium; he used sugar, ammonium, and other salts. *Today* we *can* grow yeast in a completely "mineral medium" of sugar and salts, plus the five vitamins. The vitamins, incidentally, are far more easily made in the laboratory or in the factory than is sugar. What of Pasteur's error in overlooking the vitamins present along with the yeast cells in his "pin head" seeding? It is said that Japanese artists purposely introduce a minor blemish in their finished paintings, for only God can make the perfect masterpiece. The "pinhead" then, was the insignificant blemish in the work of the man who was called by the great physician Sir William Osler, the "most perfect man who ever entered the Kingdom of Science."

The fuel of our cells

STARCH is a large, often the major, portion of man's diet. Rice, potatoes, and flour are cheaper to produce than cheese, eggs, and meat. Therefore the majority of mankind lives mainly on those three starch-laden staples. All too many get very little even of them.

Starch is but one of several different substances which the chemist groups together as sugars. The simplest sugars are grape sugar, known technically as glucose, and fruit sugar or fructose. Both of these sugar molecules contain six carbon atoms attached in a row, festooned with six oxygen and twelve hydrogen atoms. They differ in the architectural pattern of those hydrogens and oxygens.

A molecule of each of these sugars is grafted together by the sugar-cane plant to form the cane sugar with which we are all familiar. (Fructose is sweeter than cane sugar. This accounts for the great sweetness of honey; the enzymes of the bee dismember cane sugar into its components, fructose and glucose. Saccharin, the artificial sweetening agent is not a sugar at all. It is a synthetic organic molecule which, by chance, happens to have an impact on our taste buds which induces the sensation of sweetness. Saccharin is not metabolized, therefore it has no caloric value.)

Plants can also clip together hundreds of glucose molecules to form the multi-sugars—starch and cellulose. The grafting together of the many small glucose units into the huge starch or cellulose molecule is accomplished in a manner similar to the assembly of amino acids into proteins: water molecules are shed and the sugar molecules fuse at the shorn sites left by the detaching of the water.

Cellulose, which is the main component of the leaves and stems of plants, is useless to us as a food. We cannot absorb these huge molecules from our intestines; nor do we have the enzymes or the microorganisms, as the cow does, to break them up into smaller, usable molecules.

It is amusing to recall one of the many propaganda stories used by the Nazis who tried to inspire awe with their ferocious army backed by their superlative science. German paratroopers were said to be equipped with little pills, presumably of enzymes, which were supposed to enable them to live off the land, literally. They could live, we were told, on grass, like locusts. Had this been true it would have been a considerable achievement. We have not been able to concentrate, in any quantity, the enzymes which split cellulose. Only snails, wood-eating insects, and some microorganisms seem to have such enzymes. It is now possible to use microorganisms for the large-scale production of a variety of products. We have learned to grow penicillin molds in tanks of thousands of gallons capacity, multiplying by astronomical numbers the volumes that bacteriologists could handle before the Second World War, but so far we have not been able to produce cellulose-splitting enzymes this way. Of course the German army had no such cellulose-splitting enzymes.

In whatever form the sugars are eaten, starch or cane sugar, they are broken down by the juices of the alimentary canal to simple sugars, which are then absorbed into the blood stream. All the absorbed sugars are converted to glucose; that is the only sugar found circulating in the blood. This does not, however, justify the claim of some candy advertising that dextrose (another name for glucose) is the quick energy food. Cane sugar is split in the alimentary canal so rapidly that there is no difference from glucose in its availability for a normal person.

The amount of glucose in the blood is remarkably constant; it increases in diabetes, but otherwise its level is about the same in all average healthy persons.

The body's heat is derived mostly from the burning of glucose; cold-blooded animals such as the frog have less sugar in their blood than we do; birds, which are warmer than we, have more. The writer could not find out whether the blood sugar of the shrew has ever been determined. This tiniest of mammals—it is smaller than a mouse—has a body temperature even higher than that of birds. Undoubtedly its blood sugar is higher, too.

If we could not store so vital a substance as glucose in our bodies, we would have to be eating incessantly to maintain a steady supply of it. That would be a precarious existence. Should we fall asleep we would never wake; for lack of its fuel our bodies would grind to a halt. On the other hand large amounts of free glucose could not be kept in the body either; it is too readily used up. We therefore deposit glucose in a more stable, less reactive form. Scores of molecules of glucose are hooked together to form this stable reservoir called glycogen. This is the animal's version of the plant multi-sugar starch.

Glycogen is stored all over the body: there are depots of it in the liver, in the muscles, in the kidneys. The living organism husbands and distributes its resources well. As we need glucose, enough of the reserve glycogen is mobilized and is broken down into independent glucose molecules to fill the order. If there is any excess glucose in the blood, as after a meal, it is shipped to the glycogen depots. If energy is needed, the glucose molecule is broken down to release the sun's energy originally packed into it by the green plant. The release of energy is performed with an astounding series of enzyme-motivated reactions.

The earliest glimpses of these reactions were obtained, oddly enough, from studies not of animals but of yeasts. Yeast cells, up to a point, utilize sugars for *their* energy exactly as we do. (They are unable to cope with alcohol, which *we* burn with ease to carbon dioxide and water.)

The first step in the metabolism of glucose in the yeast cell or in the human cell was discovered by the same chemists who discovered coenzymes—Harden and Young. They found that yeasts starve in the midst of an abundance of glucose unless inorganic phosphate salts are present. But when phosphates are added to yeasts, they thrive on their glucose. Why the need for phosphates?

Harden and Young found that the yeasts, as the first step in fermentation, hang two molecules of phosphate on the first and sixth carbon atoms of the glucose molecule. Later, other chemists were able to cajole out molecules containing two or three carbons with phosphate still hooked onto them. An example is phosphopyruvic acid, a compound containing three carbon atoms, which upon losing the phosphate becomes pyruvic acid, the substance which accumulates in the blood of patients suffering from beriberi.

These two- and three-carbon-containing fragments of glucose are found in yeast cells and in elephant cells. All of these cells crumble the glucose gradually into smaller fragments and thus release the energy in the glucose slowly. How is this energy used by the cell? Before we can be qualified engineers for this most marvelous of machines we must learn the ABC of the energy of chemical reactions.

All forms of energy are interchangeable: heat can be converted into motion; motion can be converted into electricity, which, in turn, can give light or heat again. Such conversions are often very inefficient. The best steam engine loses about half of the energy of the steam as it converts it to motion.

Energy cannot be destroyed; we can not circumvent its complete liberation by using different paths for its release. We can take a pound of coal and burn it in ample air to carbon dioxide. An amount of heat (X) will be liberated. If we burn another pound of coal in a limited supply of air it will form carbon monoxide, but only about one fourth as much heat will be liberated as before. But if we now burn all of this carbon monoxide to carbon dioxide we get the rest of the original amount of heat (X). Whether we release the energy in one step or in several steps, the overall amount is the same.

Now let us take inventory of the energy in glucose. Computations of energy are always based on the chemist's unit weight, or molecular weight. In the case of glucose, this is 180 grams. (One molecule of glucose weighs 180 times as much as one atom of hydrogen.) When a green plant makes glucose from carbon dioxide and water it packs energy into it. Into 180 grams of glucose—about six ounces—are packed 700 Calories [1] of energy. If we burn in a stove the six ounces of glucose we will release 700 Calories of heat.

Now, where are these 700 Calories hidden? They are used to form the bonds that hold together the six carbons, twelve hydrogens, and six oxygens of the glucose molecule. There is energy in chemical bonds. Imagine a dozen large springs from a mattress squeezed into a hat box. A good deal of energy had to be expended to squeeze the springs together before the box lid could be safely locked. If the lid is opened, the jumping springs will release the same amount of energy that was used to put them into the box. This is a fair analogy of the energy used to lash the atoms of carbon, hydrogen, and oxygen together to form glucose. (This bond energy has nothing to do with the energy within the nucleus of the atom, the notorious atomic energy. And of course bond energy is dwarfed by the monstrous energy of the nucleus.)

The amount of energy in each chemical bond is not the same; some bonds have more energy packed into them than others. The cell gets its warmth and its energy for work from the breaking of the bonds of glucose. Cells are equipped with an elaborate set of enzymes for the step-wise whittling of the bonds of the glucose molecule and the stepwise release of energy from these bonds.

Some of these steps are well known, and the overall picture is clear. The bonds in glucose are broken one at a time, and phosphates play a stellar role in the process. They store much of the released energy in a form more

[1] The Calorie is a measure of heat energy. One hundred Calories will heat one liter (about a quart) of ice-cold water to boiling.

convenient for the cell. The howling wind has a lot of energy. The farmer's windmill catches some of that energy and at once puts it to work pumping water. But some of that energy is also stored by the charging of batteries. The wind can not light up the farmer's house, but the battery can. The heat from a crumbling glucose molecule can not, by itself, make our legs move but the energy in a phosphate bond can. Phosphate bonds are our batteries. They are the stored energy for life's every need.

What is the mechanism of this marvelous battery to which we owe our lives? The battery is a molecule—a molecule called adenosine triphosphate—abbreviated, ATP. Onto a molecule of ATP are lashed two special phosphate groups. Ten Calories of energy are packed into each of those bonds which secure these phosphates to the ATP. These phosphate-cementing Calories are the only form of energy the cell can use for its many tasks.

Fifty such phosphate bonds are formed from the energy released by one molecule of glucose. Fifty times ten Calories are stored from the 700 contained in the six ounces of glucose.

The other 200 Calories which are not captured into phosphate bonds keep us warm. But 500 out of 700 or 70 percent of the Calories are saved for future work. There is a loss of only 30 percent of the total energy in this transformation. The cell is thus a better engine than the best steam engine, which is only 50 percent efficient in such a conversion.

This stored phosphate-bond energy is used in an ingenious manner. Suppose the cell is in need of a substance the assembly of which requires 30 Calories of energy. Three units of ATP are alerted to act as coenzymes in the cell's assembly line; each ATP unit splits off one phosphate unit and they thus deliver the requisite 30 Calories. The three shorn ATP units in turn require replenishment of their lost energy. Glucose is mobilized from a glycogen depot and begins to be degraded. The energy flowing from the crumbling glucose is used as a cement by the three ATP units to reattach their three phosphates. Then the glucose whittling stops; the three ATPs are ready for any new emergency.

How does phosphate-bond energy move our legs? The bones in our legs are moved by the muscles attached to them. These muscles always come in matched pairs. As one muscle contracts, its opposite relaxes; then the other one contracts, and the first one stretches. We move by such a sequence of alternate relaxing and contracting. All the work in this process is done by the contracting muscle; it moves the bone and stretches its opposite muscle. The energy for this work is provided by the phosphate-bond energy of ATP. The energy for the phosphate bond comes from glucose, and this energy in turn comes from the sun. So we are sun machines like the multivaned toy in the optometrist's window—fantastically complex sun machines, but sun machines nonetheless.

Let us lift the hood and take a look at the machinery. Muscle is made up of long fibers composed chiefly of a protein called myosin. These fibers are teeming with ATP. Recently, Albert von Szent-Györgyi, a Hungarian biochemist who had already received the Nobel Prize for his earlier work, which included the isolation of vitamin C,[2] achieved

---

[2] Szent-Györgyi is the perpetrator of a celebrated biochemical pun. He isolated an unknown substance which later proved to be vitamin C. He soon learned that he had on hand a sugar which had not been described previously. As the discoverer of a new substance, he had the right to name it. The names of all sugars must end with -ose, as in glucose, sucrose, and fructose. Since he had a sugar of unknown structure he submitted the

muscular contraction in a test tube. He managed to free the myosin threads of all of their ATP. He then added to these threads, now elongated, some ATP. The long threads curled up instantly on contact with the source of energy. Contraction of myosin thread is contraction of a muscle. This demonstration is a milestone in the history of science, for in Szent-Györgyi's words: "Motion is one of the most basic biological phenomena and has always been looked upon as the index of life. Now we could produce it in a test tube with constituents of the cell."

The availability of the stored energy of ATP for movement is marvelously useful to the animal in emergencies. Sometimes the animal needs enormous amounts of energy when there is insufficient time to metabolize glucose. After a vigorous sprint to the bus one may pant for minutes before one can settle down to the calm reading of the morning paper. The panting is a forced intake of large amounts of oxygen needed for the burning of glucose to replenish the ATP used up by the exertion of hurrying to the bus.

The stretching of muscle threads lacking ATP explains the hitherto puzzling rigidity (*rigor mortis*), which sets in soon after death. The ATP present in the muscle is slowly decomposed after death, and, since the enzymes of sugar metabolism are forever stalled, the ATP is never reconstituted. Without ATP, the muscle fibers stretch and cast the corpse into the rigidity of death.

ATP has also been shown recently to be the source of energy for the light which some organisms are able to pro-

name "ignose" to the British scientific journal *Nature*. The editors of that journal frowned on such frivolity and asked for a new name. "God-knows" was Szent-Györgyi's prompt reply.

duce. It is also the source of electrical energy in nerve tissues and in the electric organs of animals which can accumulate and discharge electricity. Thus, glucose is the fuel of the cell, but the energy flowing from it is stored in this most versatile of reservoirs, ATP, which can be tapped to supply all the forms of energy a living organism can generate: heat, light, mechanical, electrical, and chemical energy. With these recent findings the biochemist of the twentieth century has completed the evidence for the mechanistic concept of life which the biologists of the nineteenth century so enthusiastically espoused.

The writer has not, as yet, mentioned insulin, which, as everyone knows, is essential in sugar metabolism.

How does insulin fit into that elaborate maze? An astonishing amount of work went into the winning of insulin. This work illustrates the effort needed to expose and understand the role of but one small cog in the complex machinery of the cell.

Diabetes is one of man's worst scourges. There are two different diseases which bear the same name, diabetes mellitus and diabetes insipidus. The only similarity between the two diseases is the same distressing symptom, the passing of enormous volumes of urine. The word diabetes actually means a siphon and the qualifying adjectives, mellitus and insipidus are remnants of the days in medical diagnosis when, unaided by chemistry, the hapless physician was forced to differentiate between the two diseases by the taste of the patient's urine, pronouncing the disease either mellitus (sweet) or insipidus (tasteless). Differentiating the two diseases is about all the physician could do until the early 1920s, when insulin was given to the grateful medical profession and, of course, the even more grateful patients.

Most of the early work in this field was done by physiologists, biological scientists whose interest is the function of cells and organs. They very quickly discovered one of the functions of the pancreas, an organ found just under the stomach. The pancreas produces solutions of enzymes which are poured through ducts into the small intestine, where they split proteins, fats, and sugars.

That the pancreas has other functions, too, was suggested as far back as 1686 by a physician named Johann Conrad Brunner who thought that the pancreas was in some way involved in the utilization of fats and sugars. Two hundred years later there was complete confirmation of this hunch.

In 1889 two physiologists, Oscar Minkowski and Joseph von Mering removed, under anesthesia, the pancreas of dogs. The dogs survived the operation, but in four to six hours began to show the characteristic symptom of diabetes mellitus; they were voiding sugar in their urine. As much as two ounces of sugar was lost by one dog in a day. At the same time the sugar in the dogs' blood increased. They became bona fide diabetics.[3]

Prior to these operations it had been known that the duct leading from the pancreas to the small intestine could be completely blocked without harming the dogs in any way. Apparently dogs could do without some of the products of the pancreas: the digestive enzymes which are poured through the ducts. The enzymes of the stomach enabled

[3] The probably apocryphal story is told that it was the caretaker of the dogs who discovered the sugar in their urine. He is supposed to have noticed that swarms of bees followed the depancreatized dogs.

them to hobble along. The pancreas must, therefore, exert its influence on sugar utilization through some medium other than the juices pumped through the ducts into the intestine.

Complete confirmation of this dual role was provided by a brilliant operation of Minkowski. He removed the whole pancreas of dogs but he immediately grafted pieces of their pancreas under their skin. The dogs survived such operations and led a fairly normal, undiabetic life. With the pieces of pancreas inserted under their skin, there was no possibility of any enzymes getting into their intestines. Whatever the pancreas produced must have gone directly into the blood stream of the dogs.

Insulin was the fruit eventually harvested from these early experiments. Millions of diabetics of this and of yet unborn generations owe their painless days to these discoveries.

They were remarkably fortunate discoveries. Minkowski tried to repeat the production of diabetes in other experimental animals. He was unsuccessful with pigs, goats, ducks, and geese. Finally, with the cat, he was able to duplicate his earlier discovery on dogs. These two are the only experimental animals which develop a positive, unequivocal picture of diabetes on the removal of the pancreas.

The next great stride forward was made by a young student, Paul Langerhans, who was studying for his doctoral dissertation the structure of the pancreas. He cut thin sections of the organ and saw with his microscope two entirely different types of cells. There were grapelike bunches of cells, and among these there were small islands of cells which were different in appearance.

As early as 1893 it was suggested that it is the island cells—the islands of Langerhans as they came to be called—that produce something which is essential for the normal handling of sugars. There was confirmation of the relation between the island cells and diabetes from the examination of the pancreas of dead human diabetics. Invariably such post-mortem examinations revealed unnatural-looking, degenerated island cells.

There were immediate attempts to apply the newly found relationship between the island cells and diabetes to a possible cure of the dread disease. The spur was its wide prevalence—it afflicts about one percent of our population—and its relentless course through emaciation, muscular weakness, and final infection, to death in a few years.

The feeding of organs of healthy animals to patients with diseased organs was an ancient art and superstition. Minkowski himself was the first to try the feeding of the healthy pancreas of other animals to his depancreatized dogs. He obtained no improvement whatever. But the hope that some active principle might be extracted from a normal pancreas spurred on workers for the next thirty years.

This period was by no means fruitless. Chemists developed accurate methods to assay the sugar in the blood of animals in samples as small as a drop. Scores of physiologists plugged away at the preparation of extracts, all of which, alas, turned out to be toxic or impotent, or both, when injected into depancreatized animals.

Many were within a hair's breadth of reaching the solution, but history and the Nobel Prize Committee remember only the one who closes that final small gap. Frederick G. Banting of the University of Toronto has been acclaimed as the discoverer of insulin, and has received many honors including knighthood and the Nobel Prize. But we must remember that insulin was not the product of the "flash of genius" of one mind. Scores of scientists from Minkowski on have accumulated new knowledge, and from this pooled information arose a pattern, like a laborious jigsaw puzzle, lacking only the final fragment of information.

The futile question is sometimes asked: What if there had been no Banting? The answer is: Someone else would have extracted insulin successfully somewhat later. There were others who were following the same hypothesis and almost the same procedures as Banting.

This does not imply that there is no genius among experimental scientists. It merely means a different manifestation of genius; in the field of science, genius accomplishes what lesser minds would accomplish later. Creation in the arts is quite different. It is inconceivable that anyone but Shakespeare or Beethoven might have brought forth those very same plays and symphonies. But Newton and Leibnitz independently and almost simultaneously integrated the same mathematical abstractions into differential calculus. The artist extracts his creation almost solely from the riches of his own mind; the scientist evolves in his mind a pattern from phenomena which he and others have pried out from observations of our physical universe. Genius among scientists can be measured in years—the number of years that he is ahead of his contemporaries.

In 1920 Banting, a twenty-nine-year-old Canadian physician, read an article on surgery. It was a description of the effects of the blocking of the ducts leading from the pancreas into the small intestine. The survival of the island cells amidst the degenerating grape cells was emphasized. A brilliant idea took shape in Banting's mind. Heretofore everyone had ground together the whole pancreas in the first step of the preparation of insulin (that name had already been given to the long-sought, active agent of the islands of Langerhans). Was it not possible that the failure to extract an active preparation was due to the destruction of insulin by the enzymes of the grape cells? These grape cells were bursting with potent enzymes which gushed out on grinding. This article in surgery held the answer. Tie off the pancreas; let the grape cells wither and then try to extract insulin from the intact island cells. Lacking the facilities for carrying out this project, Banting applied to the department of physiology of the University of Toronto for help. He was joined by a young physiologist, Charles H. Best, and the two started the quest for the elusive insulin. They tied off the pancreas of several dogs, using standard surgical techniques and care on their patients. After about two months they sacrificed the dogs, removed the pancreas from each, froze these organs, and ground them up in a salt solution which simulates the salt content of the blood.

This mash was filtered free of insoluble debris and the clear solution was injected into a vein of another dog whose pancreas had been previously removed and which by then was in the advanced stages of diabetes. After the injections of these crude preparations the blood sugar of the depancreatized dog was lowered. Insulin was born!

The extraction and purification of insulin could proceed on a large scale once it was demonstrated that the enzymes of the grape cells were the enemies to be thwarted. These enzymes were known to be inhibited by acid. Therefore the whole pancreas glands of cattle were extracted in acid solutions and the extracts were purified somewhat to make them nontoxic. Such an extract was first used on a fourteen-year-old boy in the Toronto General Hospital, who was suffering from severe, hopeless diabetes. His blood sugar was immediately reduced by 25 percent. He was the first of millions to benefit from this new weapon in our all too meager arsenal against disease.

What is insulin and what does it do in the cell? We know the answer to the first question; to the second there is only a partial answer. Insulin was obtained in pure crystalline form and it turned out to be a protein. This explains why feeding whole pancreas or insulin to patients is useless; also why it was impossible to extract it until the enzymes of the grape cells were thwarted. In both cases insulin is destroyed by the protein-splitting enzymes—in the first case by the enzymes in the patient's stomach, in the second by the same enzymes oozing out of the grape cells.

Insulin is poured directly into the blood stream by the island cells. It is but one of a number of substances produced by ductless cell clumps, or glands, which regulate the activity of other cells. The fruits of these ductless glands are called hormones—a word derived from the Greek verb meaning "to rouse to activity."

The preparation of insulin has become a major industry. The pancreas glands of cattle are shipped from the slaughterhouses to pharmaceutical plants for processing. That is why a dish of sweetbreads will most likely be not sweetbreads, the culinary name for pancreas, but thymus, another gland from which the physiologist and the biochemist have not as yet been able to extract anything of value.

The potency of insulin preparations was standardized by the health organization of the League of Nations. The signal success of the international standardization of many drugs and vitamins by this organization contrasts starkly with its failures in the political field. We must not ascribe these successes wholly to the absence of politicians and diplomats from the committees in charge of this work; it is a little easier to agree on standards of insulin, than on sanctions against Italy.

The exact, complete role of insulin in the utilization of sugar is not yet known. There are recent fragments of information, indicating that it exerts its profound influence by accelerating some of the enzymes involved in the breakdown of sugar. It is surprising what a small cog insulin is in the complex machinery of sugar metabolism, in which dozens of different enzymes have already been identified. But the cell's machinery is adjusted with exquisite delicacy. The slightest imbalance at any one point may pile up sugar and flood the diabetic patient with it, wreaking, like all floods, havoc in its path.

It is sobering to speculate on what our knowledge of diabetes or of insulin would be today if Minkowski and Mering and the physiologists and biochemists who followed them had been forbidden to use cats and dogs, or, for that matter, any experimental animals. Such speculation is not idle for there have always been a small number of people who have banded together into antivivisection societies whose main function is agitation for the proscription of all animal experimentation. The motivation of antivivisectionists is usually a squeamish humaneness which blinds them to the benefits of experiments on animals: the mastery of diseases which plague not only them, but their very pets.

It is one of the strengths and beauties of a free society

that little bands of people like this, can bob up and down on its outer fringes unmolested, while the main stream of society flows on majestically, unconcerned by the little bands which are frantically pouring out a variety of pamphlets, calling on the main stream to be diverted or dammed.

But lately the antivivisection movement has taken on a new aspect. It has become a streamlined, hard-hitting, high-pressure lobby, squeezing every state legislature and Congress to enact laws forbidding animal experimentation. It seems to have abundant funds, is backed by some newspapers, and publishes testimonials from Hollywood stars on the futility and cruelty of animal experiments.

Could the antivivisectionists have their way, not only would all research cease, but a great many drugs now available could not be used, for, to insure their safety, they must be tested on experimental animals. Of the 600 drugs listed in the Pharmacopoeia of the United States, our most authentic authority in this field, 274 either were originally developed through animal research or must be tested, as each new batch is produced, on animals, to insure their safety for human use.

Close to sixty people were killed a few years ago by a new drug preparation which had not been tested on animals! These corpses, robbed of their spark of life by an "Elixir of Sulfanilamide," are testimony to the absolute necessity of testing a new preparation on animals. Sulfanilamide is quite insoluble in water and therefore cannot be dispensed in a convenient aqueous solution. An obscure pharmaceutical manufacturer tried other liquids as a solvent. He hit on ethylene glycol, and blithely sold such solutions of sulfanilamide to be taken by the spoonful. Unfortunately, ethylene glycol is converted in the body to oxalic acid, a potent poison.

Had the preparation been tested on just one dog first, those people would not have gone to their early graves. What human wealth might have been saved by the life of just one dog. Just one dog, spared from the hundreds of thousands destroyed annually at dog pounds, to test the "Elixir of Sulfanilamide." In the year of this disaster, in New York city alone, 55,000 stray dogs and 150,000 cats were destroyed at the pounds by asphyxiation!

5 Isotopes

$$CO_2 \quad CO_2 \quad CO_2$$
$$CO_2 \quad CO_2 \quad CO_2 \quad CO_2$$
$$CO_2$$

Tracers for exploring the cells

THAT FOODS are "burned" in our bodies to carbon dioxide has been known since 1789. We are indebted for the knowledge to the great French chemist Antoine Laurent Lavoisier, who founded modern chemistry and, in a limited sense, biochemistry as well. He boldly declared: "La vie est une fonction chimique."

Lavoisier elevated chemistry to a science by abandoning speculation on the nature of chemical reactions in favor of observation and measurement. After exploring the nature of combustion of inanimate objects he undertook a similar study of "combustion" in living organisms. With astonishing experimental skill and judgment he showed that the process was the same in both the living and nonliving worlds. A guinea pig and a burning candle both produce carbon dioxide. Furthermore, he measured the amount of heat liberated by the candle and the animal and showed that the heat in each case was proportional to the amount of carbon dioxide produced. (He overlooked the heat liberated by the production of water.)

This genius was ordered to trial by the National Convention and was guillotined in 1794. He was denounced by Marat, whose own ambitions as a scientist had been thwarted by him. Marat had published his own theory on the nature of combustion which was essentially a rehash of some stale alchemistic concepts. He denied that oxygen has a role in the process. Lavoisier had proved the views of Marat, the would-be scientist, wrong; but Marat, the politician and rabble rouser, unfortunately had the last word. After the revolution began, Lavoisier became a constant target of vituperative attacks in Marat's newspaper. His crime was that he had been a farmer-general of France. Although during his tenure of office this notorious tax-collecting agency had undergone many reforms, Lavoisier was nevertheless accused and convicted of guilt by association with the once corrupt agency. Testimonials to his great service as a scientist to France and to the cause of the Revolution—he was in charge of standardizing weights and measures—were of no avail. It is reputed that Lavoisier requested a two weeks' stay of sentence so that he might finish some experiments on respiration, but the presiding judge, one Pierre Coffinhal, is said to have replied: "The Republic has no need of scientists; justice must follow its course."

For a long time after Lavoisier's death, life was compared to a burning candle. However, this analogy could not be maintained indefinitely, for obviously a living organism, unlike a candle, is not consumed in its own flame. After the combustion engine was invented, the living organism was compared to *that*. (Man often belittles the grandeur of life's machinery by comparing it to one of his own handiworks. These days it is the vogue to compare the brain to an electronic computing machine.)

For a long time food was thought to be merely a fuel for the machine of the body. That the quality of the "fuel" was as important as its quantity did not necessarily conflict with the image of the body as a combustion engine.

However, it slowly became apparent that the living machine is a most unusual one: the parts of the machinery themselves seem to be burning as well as the fuel. The great American biochemist Otto Folin, who was appointed, in 1906, the first professor of biological chemistry at Harvard, demonstrated that tissues, too, are broken down independently of the breakdown of foods. He found the excretion of certain waste products from our bodies to be unaffected by the amount or kind of food. He attributed to the metabolism of the tissues themselves these constant waste products.

To what extent, if any, the components of the diet interacted with the tissues could not be determined. For, once the food enters the blood stream—after digestion in the alimentary canal—it disappears from sight. It becomes hopelessly intermingled with substances similar to it that are already present in the tissues. Until recently, it was impossible to distinguish such molecules, which were already components of the tissues, from molecules only recently absorbed from the digestive tract.

For example, what happens to a pat of butter we eat? Before we can trace the path of that butter in the body we must first become acquainted with a little of the chemistry of butter. Butter is a fat. A typical fat molecule is built from three molecules of fatty acid and one of glycerol (the glycerin of the pharmacist). Fatty acids are made principally of carbon and hydrogen. The carbon atoms are strung together like beads with two hydrogen atoms linked to each carbon bead. The last carbon of the chain has no hydrogens. Instead, two oxygen atoms are attached to it. This aggregate of three atoms is the acidic group. The chains of carbon vary in length; the shortest is acetic acid (present in vinegar), made of two carbons; the next longer one (present in fats), contains four carbon atoms, the next six, and so they proceed with two carbon increments, up to twenty-four. There is always an even number of carbons, never an odd number, in any natural fatty acid. A natural fat such as butter contains a variety of fatty acids, short and long.

By means of their three acidic groups, which may be likened to "hooks" three fatty acids are attached to a molecule of glycerol, which contains three appropriate "eyes." The coupling between fatty acids and glycerol takes place, as with amino acids, by shedding water. The product made by the coupling of three fatty acids and of glycerol is the fat molecule.[1]

---

[1] We must marvel at the ingenuity with which nature employs a simple process. To make proteins she splits out water, to make starch or cellulose she splits out water, and now she does the same for the shaping of the fat molecule. We must remember that life started in water and continues in water. We are indissolubly wedded to water because the surface of the planet we happen to inhabit abounds in that fluid. We are "creatures of circumstance." If our earth had an average temperature a few degrees lower, the story might have been different. Water, instead of being an abundant liquid, would have been a relatively high melting mineral, ice. (Even now much of the earth is covered with ice.) In that case, living creatures, if any could have arisen, might have had, instead of water as the major constituent of their tissues, another fluid—perhaps liquid ammonia. Chemical reactions within the bodies of such nonaqueous creatures would not involve the splitting out or addition of water. They would use ammonia as the chemical zipper in the assembly or breakdown of *their* large molecules.

If we cook a fat with hot alkali, we dismember the fat molecule and convert it back into its components: the fatty acids and glycerol. The fatty acids combine with the alkali to form a soap. This operation is the basis of the tremendous soap industry. Soapmaking has been practiced essentially the same way for thousands of years. Any available fat— beef tallow or a vegetable oil—has been cooked with whatever alkali was available. The pioneers in the American wilderness leached out the alkali ashes of burnt wood; today we use lye produced with electrical energy from table salt. Fundamentally all soaps are the same. They differ in color and odor, and in the amount of air, salt, and water put into them.

Butter is broken down by enzymes, mostly in our small intestine, into fatty acids and glycerol. The enzymes do, with cool efficiency, what the alkali does in the hot cauldron. The enzyme-produced fragments are absorbed from the intestine and disappear. Where they go and what they do was a complete mystery until a few years ago. Of course we knew that eventually they must be crumbled down to carbon dioxide and water, but nothing more was certain.

Are they disintegrated immediately or do they linger in the body for a while? If they linger, where are they? Do they go to the liver, or do they go to the blobs of fat depots which form all too readily around our waistlines? Do they fall apart into carbon dioxide and water at once or do they disintegrate gradually? Do the different fatty acids have the same nutritional value?

These are the questions which challenged the biochemists' imaginations. They appeared to be questions doomed to dangle before us without answers. For once the fragments of the fat molecule are absorbed from the intestine they become hopelessly intermingled with the pool of other fat molecules already present in the body and become indistinguishable from them.

There have been attempts to label a fat, to hang a bell on it, in order to chart its wandering throughout the body. In one such attempt some of the hydrogens in a fat were replaced by entirely different atoms of another element, bromine. The strategy of this experiment was simple. The presence of bromine is very easy to detect and normally the amount of bromine in tissues is very, very small. If after eating the bromine-labeled fat, a rat's liver should contain a large amount of bromine that would be an indication that the fat entered the liver from the intestine.

While the scheme sounds simple and effective on the surface, actually, labeling a fat with bromine yielded no worthwhile information. Such a drastically altered fat, in which bromine atoms had replaced hydrogen, is an unphysiological substance. The enzymes of the body, which are notoriously fastidious in the choice of substances on which they act, may have nothing to do with such unnatural substances. A label which could go unnoticed by enzymes was needed. Such a label for the fat molecule was provided, not by biochemists, but by atomic physicists. With the aid of this label we not only tracked down the fat molecule but also learned so much that we changed our whole concept of the mechanism of the cell.

Since isotopes have been seared into our minds by the heat released over Hiroshima and Nagasaki only a brief summary of them need be presented here. (If the reader is interested in the exploratory adventures of the mind which blazed the trail to Hiroshima and Nagasaki he should read *Explaining the Atom* by the late Selig Hecht who presented the drama, without histrionics, of this triumph of the human mind.)

The atoms which compose an element are not uniform in weight. The vast majority of hydrogen atoms have the same weight. But there is one atom in every 7,000 which weighs twice as much as one of its more abundant lightweight brothers.

In the case of nitrogen most of the atoms are 14 times as heavy as the common hydrogen atom. (Their atomic weight is 14.) But one nitrogen atom in 270 is 15 times as heavy. (Its atomic weight is 15.) The atomic brothers of different weights are called isotopes. They are identical twins in all respects except their weight.

A light hydrogen atom is composed of a nucleus of a single, positively charged speck of matter, a proton. Rotating rapidly around this proton is a still smaller speck—a negatively charged electron. A heavy hydrogen atom also has only one satellite electron, but its nucleus is different: it contains, in addition to a proton, a neutron—a particle almost equal in weight to a proton, but without a charge on it.

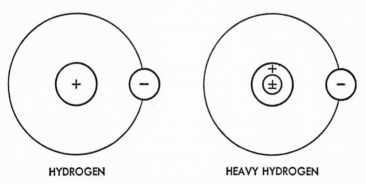

HYDROGEN   HEAVY HYDROGEN

In this age of specialization the origin of these different isotopes is not the concern of the biochemist. He lets the atomic physicist worry about them and write popular books on them. Since isotopes differ only in their nuclei, which are not involved in ordinary chemical reactions, their chemical behavior is identical. We know that the heavy and light isotopes enter into ordinary chemical reactions exactly the same way. But what about chemical reactions within the cell? Do the ultrafastidious enzymes differentiate between isotopes? There is a way to decide this question. If the enzymes which build up the body tissues discriminate between the isotopes, one or the other of two isotopes should be more concentrated in the cell than in the inorganic world.

Let us examine the nitrogen atoms which are built into the proteins of our tissues. Let us choose an expendable tissue such as hair for our studies. With appropriate chemical manipulations we can obtain billions of atoms of pure nitrogen from a few strands of hair. Where did these atoms of nitrogen come from? Originally they came from the atmosphere. But before they entered our body, the nitrogen atoms had sojourned in the bodies of many different plants and animals and therefore they had participated in a multitude of enzyme-motivated reactions.

Eighty percent of the atmosphere is nitrogen containing the two isotopic varieties, nitrogen 14 and nitrogen 15, (Both of these varieties of nitrogen are stable; unlike the uranium isotopes, they are not radioactive.) Plants and animals are unable to use nitrogen gas from the atmosphere directly. They lack the enzymes for this task. Certain bacteria which grow on the roots of legumes have such enzymes to incorporate the nitrogen from the atmosphere into their cells. From such bacteria a clover might absorb the nitrogen atoms destined for the hair of a human being. If the clover is eaten, the nitrogen, now incorporated into amino acids, would pass into an animal. The nitrogen might be returned to the soil either by excretion or upon the death of the animal. From the soil another plant might absorb the nitrogen. (In this so-called "fixed" form, the nitrogen can be utilized by plants without the aid of nitrogen-fixing bacteria.) Thus, the nitrogen atoms which are now incorporated in human hair may have sojourned in the cells of hundreds of different plants and animals; but in all the multitude of chemical reactions inside a vast variety of different cells the nitrogen isotope was neither concentrated nor diluted. For the proportion of heavy isotope in the nitrogen obtained from hair is *exactly* the same as that of the heavy nitrogen isotope in the atmosphere.

The enzymes of the cell cannot tell the isotopes apart. But the physical chemist, with his instruments, can. Now we had our label. Now we had the bell to hang on the fat molecule.

The first scientist to use an isotopic tracer in biology was George Hevesy, a young Hungarian physicist working in England. In 1923 he immersed the roots of bean plants in solutions containing a radioactive isotope of lead. He traced the ascent of the lead into the stem and the leaves of the plant by simply measuring their radioactivity with appropriate electrical instruments. Of course, since there is no lead, only carbon, hydrogen, and oxygen, in a fat, Hevesy's studies were of no help in the study of the wanderings of fat in the animal. But his success pointed the way to the use of similar "tracers" should they become available.

The labeling of a fat had to wait more than ten years until Harold C. Urey separated the heavy isotope of hydrogen—deuterium—from the light ones.

The pioneers in the use of deuterium in the study of the overall fate of foodstuffs—or their metabolism—were Rudolph Schoenheimer and David Rittenberg. As the isotopes became more readily available other research teams were formed until today there is hardly a biochemical laboratory which does not use isotopes as a tool.

To carry out an investigation using isotopes was no easy task. First of all the test compound enriched with isotope had to be prepared. A fat was made which contained not an ordinary fatty acid, but a fatty acid in which an abnormally high number of the atoms were replaced by the heavy isotope of hydrogen—deuterium. This deuterium-containing fat was fed to rats kept in suitable cages so that their urine and feces could be easily collected.

To the amazement of every biochemist and physiologist only a small fraction of the deuterium appeared in the excreta of the animals the next day. Apparently the dietary fat is not burnt up immediately in the body.

Where was the newly eaten fat? Where was the deuterium stored?

Rats were killed one, two, three, and four days after the feeding of the original deuterium-labeled fat. The various organs—the liver, the brain, as well as blobs of abdominal fat were separately cooked with alkali. The released fatty acids from each were isolated and the deuterium in them was determined. Most of the deuterium, therefore most of the original fat, was in the fat depots.

The first intelligence gained from our isotopic detection then, is that most of the fat from the diet enters the fatty depots first. How long does the fat stay in the fatty depots? The longer the rat lived after the original isotopic meal (they were on a normal rat diet after that), the less deuterium remained in its body. In three days one half of the deuterium, therefore one half of the newly acquired fat, had been used up from the fat depots. In other words, the fat which is incorporated into the body after a meal is used up only slowly. But the animals were not losing weight, therefore as the depot fat was used up, it must have been replaced from more recent meals.

This new intelligence revolutionized our whole concept of life's economy. Previously it had been thought that entering food was immediately burned up. Only the excess over the body's daily requirement was believed to be stored in depots. These depots, in turn, were believed to be inactive reservoirs tapped only on lean days. The living organism had been visualized as a combustion engine, receiving its fuel—the food—and converting it into energy and waste products without any alteration in the structure of the engine.

But the isotopes told a different story. The information obtained from studies of the fats, later confirmed with other foods, proved that the body is in a constant state of flux. Its tissues are being built up and broken down simultaneously. The molecules which compose our bodies today will be gone in a few days and replaced by new ones from our foods.

The body as a combustion engine is approximated by the train on which the Marx brothers were once escaping from one of their dire predicaments. For want of fuel in the coal car they tore up the coaches, feeding the wooden planks into the engine. Had they been repairing the coaches at the same time from fresh supplies of lumber they would have almost simulated the engine of the body.

Rudolph Schoenheimer, whose brilliance as an investigator was matched by a rare talent for vivid exposition offered a regiment as an analogy for a living organism: "A body of this type resembles a living adult organism in more than one respect. Its size fluctuates only within narrow limits, and it has a well-defined, highly organized structure. On the other hand, the individuals of which it is composed are continually changing. Men join up, are transferred from post to post, are promoted or broken, and ultimately leave after varying lengths of service. The incoming and outgoing streams of men are numerically equal, but they differ in composition. The recruits may be likened to the diet; the retirement and death correspond to excretion." He added, however, that this analogy is "necessarily imperfect."

It is impossible to evoke a perfect analogy to a living organism. It is trite but true that the only analogy to a living organism is another living organism.

Another bounty from the study of fats with the aid of isotopes is the solution of an old riddle in the metabolism of fats. It had been known for a long time that animals can convert sugar or starch into fats. An animal receiving but a small amount of fat in its diet along with large amounts of starch produces huge fatty depots far in excess of the total amount of fat eaten. The fattening of cattle on a diet of corn, which is high in starch and low in fat content, is a practical evidence of this transformation. The corn starch is converted by the cow's cellular alchemy into the characteristic firm streaks of fat in a good cut of beef. A goose must suffer a fatty enlargement of its liver to produce a *paté de foie gras* to a gourmet's taste. It is forcibly fed huge amounts of starchy meals; corn is shoved mercilessly down its unwilling throat. It produces so much fat that its liver becomes gross and gorged with fat.

To see whether animals can really make all of their fats, an interesting experiment had been carried out almost twenty years ago by a husband and wife biochemical team, George O. and Mildred M. Burr. Rats were kept on a diet completely devoid of all fats. Such a diet contains about 25 percent of casein from which all fats are removed with ether, which dissolves the fat but not the protein. The diet also contains over 70 percent of cane sugar. Commercial cane sugar, which is so pure it contains nothing else, is used. The rest of the diet consists of a salt mixture and all the vitamins.

At first rats do quite well on such a diet. But after a while it gradually becomes apparent that something is wrong; the rats do not put on their daily quota of weight. In about seventy days the rats are a sickly looking lot. Their tails are scaly, ridged, and fragile; pieces of the tails crack off. Their hair is full of dandruff and falls out in clumps. That their internal organs are also damaged is obvious from their bloody urine; the kidneys must be cracking too. If the rats' diet is not changed, these kidney lesions will kill the animals. But if before the rats reach a moribund state a few drops of fat—lard or a vegetable oil—are given to them daily, they miraculously recover.[2]

What is there in a fat which protects the rats against this loathsome disease? Besides their varying lengths, fatty acids differ in another respect. They do not all contain their

---

[2] There are recent reports from England indicating that humans, too, can suffer from this fat deficiency. From the start of the Second World War the English diet has been lacking in fats. Scaliness of the skin which has become more and more prevalent, especially among older people, is believed to be caused by this deficiency.

full complement of hydrogen atoms. In a normal, or saturated, fatty acid there are two hydrogens for each carbon atom. However, in some fatty acids there are carbon
atoms with only one hydrogen allocated to them. These
are the so-called unsaturated fatty acids. Usually a fat contains a mixture of saturated and unsaturated fatty acids,
the liquid fats being richer in the unsaturated fatty acids.

The various components of a fat were fed to the rats
made sick by the lack of fats in their diet. The charm which
protected them was not glycerol; it was not the saturated
fatty acids; it was an unsaturated fatty acid called linoleic
acid.

If minute amounts of this doubly unsaturated fatty acid
are fed along with the fat-free diet, the rats lead a normal
healthy existence. Why must rats receive this fatty acid
and not the others in their diet? Isotopes gave the answer.

A group of mice on a normal diet received injections of
water in which the hydrogen atoms were replaced by their
heavy twins deuterium atoms.[3] (Such water has a greater
density than ordinary water and is therefore called heavy
water. It looks and tastes like ordinary water; only the sensitive instruments of the physical chemist can tell the two
apart.) A few days after the injection the mice were killed
and their various fatty acids separated.

The saturated fatty acids contained large amounts of deuterium. How did deuterium get into these fatty acids? It
could be there only if the mice made these fatty acids, using hydrogen from their body water to string onto the carbon skeleton. After the injection of the heavy water their
body water contained not ordinary water but deuterium-

---

[3] Heavy water was rare and expensive. Only one fifth as much heavy water
is needed for an experiment with a mouse as with a rat.

enriched heavy water. Since the enzymes do not discriminate between the two isotopes, both ordinary hydrogen and
deuterium were strung onto the carbon skeleton of the saturated fatty acids.

The charm, linoleic acid, which wards off the disease
produced by the fat-free diet, was also isolated from these
mice. It contained no deuterium at all! This is absolute
proof that the mice could not make this compound; if they
could, there would have been deuterium in it.

The isotopes taught us a new profound lesson. The animal's cells need linoleic acid for some unknown purpose.
Since they cannot make linoleic acid they depend on their
diet to supply this precious, essential substance. The animal body must have a huge variety of different substances
in its cells for their healthy, smooth functioning. The cells
can make most of these. Those they cannot make must come
from their diet or they perish. Hence the need for vitamins,
and as we shall see later, for essential amino acids and for
a few other miscellaneous substances such as linoleic acid.
(Some biochemists consider linoleic acid a vitamin.)

Some species of animals are not slaves to all of these essential substances. The rat, for example, never comes down
with scurvy. It needs no vitamin C from its diet; it can make
its own. Plants can make all of their amino acids and vitamins from salts, water, and carbon dioxide. Animals are,
in a way, parasites living on the plants. Some microorganisms are parasites, too. We differ only in the degree of para-
sitism. The yeast can do very nicely on sugar and five members of the vitamin B family. The red bread mold, *Neurospora,* needs only sugar and one vitamin, biotin.

How did animals become so nutritionally dependent?
How did they forget the know-how of vitamin making?

Neurospora, the red bread mold, answered these questions for us. The story of *Neurospora* and what has been learned from it will be related in Chapter 9.

While animals have qualitative shortcomings in their manufacturing abilities, work with isotopes revealed that prodigious synthesizing activities take place in animals as well as in plants. Plants are able to fashion a large variety of complex products from carbon dioxide—a molecule containing but one carbon atom. Using the energy of the sun, plants are able to lash together several molecules of carbon dioxide to form elaborate structures. Animals lack the ability to use the energy of the sun directly. But with the stored energy of the sun in the form of carbohydrates, or more specifically of ATP, animals, too, can synthesize complex molecules from smaller carbon fragments including carbon dioxide.

That carbon dioxide can be used for the shaping of larger molecules by other than plant cells was first shown by G. Harland Wood and C. H. Werkman of the University of Iowa. This team composed of a biochemist and a bacteriologist showed that bacterial cells, though lacking chlorophyll, can attach carbon dioxide to a three-carbon-containing molecule to form a larger molecule of four carbon atoms. Their discovery elevated carbon dioxide from the role that had been ascribed to it earlier—merely a waste product in bacterial and animal metabolism. The demonstration of the new synthetic role of carbon dioxide in bacterial cells was made by chemical means. Wood and Werkman showed by classical analytical methods the accumulation of the four-carbon-containing molecule.

The unequivocal demonstration of a similar function of

carbon dioxide in animal cells had to await isotopic tracers, for it is much more difficult to show the formation of an intermediate product in animals. The complex metabolism of animal tissues does not permit the abnormal accumulation of any such intermediate. A research team of a biochemist and a physicist, E. A. Evans and L. A. Slotin [4] of the University of Chicago, proved that animal cells can duplicate the feat of a plant cell; they too can use carbon dioxide as a building block for the assembly of larger molecules. Evans and Slotin used carbon dioxide in which the carbon atoms were a radioactive isotope. By means of a Geiger counter they were able to show that the radioactive carbon dioxide was incorporated into sugar molecules within the animal cells, for the sugar they extracted after the administration of isotopic carbon dioxide was radioactive. Such a demonstration would have been impossible without an isotopic tracer. Ordinary chemical analysis could show only a *net* increase or decrease in the sugar content of the tissues. But since the components of tissues are in a state of flux with both synthesis and degradation going on at the same time, the formation of new molecules from carbon dioxide would not necessarily result in a net increase. A corresponding number of molecules might have been degraded in the dynamic tug of war between breakdown and synthesis within the tissues.

That very extensive synthesis takes place in animal cells using two-carbon-containing fragments as the building

[4] Dr. Slotin, a Canadian physicist, was killed in a tragic accident while working on the Atom Bomb Project. At the end of the war, just before he was to return to the University of Chicago, while he was instructing his successors, a mechanical failure caused the release of intense radiation. Dr. Slotin was able to remedy the failure in time to save his colleagues who were at a distance from the source of the energy but he himself succumbed within a few days to the effects of the lethal radiation.

blocks has also been shown with the aid of isotopes. Cholesterol, a complex molecule containing 27 atoms of carbon, 46 atoms of hydrogen, and one of oxygen is made from several molecules of the two-carbon-containing fatty acid, acetic acid. (If acetic acid which contains isotopic carbon is fed to an animal its cholesterol will contain large amounts of the isotopic carbon.)

Very little is known about the metabolism of cholesterol. This complex molecule is present in every cell of the animal body; it is particularly abundant in nerve tissue. In the degenerative disease arteriosclerosis, there is a rise in the cholesterol content of the blood and its deposition in the walls of the blood vessels causes their "hardening." The vessels lose their elasticity. Although cholesterol has been known since 1775, its chemical structure was decoded only about twenty years ago. It was found that the kernel of cholesterol structure appears in a variety of hormones— the sex hormones and the hormones of the adrenal cortex, including cortisone. It had been conjectured that one of the functions of cholesterol might be to serve as the raw material for the production of some of these hormones by the appropriate glands. That the conjecture was valid was proved by a biochemist who wielded his isotopic tools with ingenuity. He prepared, by chemical methods, cholesterol with large amounts of heavy hydrogen in the molecule. If, after feeding such labeled cholesterol, some of the hormones of an experimental animal contained heavy hydrogen it would clearly demonstrate that the precursor of the hormone was cholesterol. The choice of the appropriate experimental animal was crucial for the success of the experiment. Since they are so potent, hormones are made only in minute quantities in the animal glands. Feeding the labeled cholesterol to a small experimental animal and then attempting to isolate a hormone from one of its glands would have been futile: the amount of hormone present is too small for successful chemical manipulation. Feeding the precious cholesterol to a large animal such as a cow or a horse would have been prohibitive. However, it was known that some of the sex hormones are normally excreted through the urine, but only in minute quantities (12,000 quarts of urine yield 10 milligrams of the male sex hormone). But, it was also known that pregnant animals excrete one of the sex hormones—pregnanediol—in somewhat larger quantities. The labeled cholesterol was fed to a pregnant woman and from her urine the pregnanediol was isolated.[5] It contained considerable amounts of heavy hydrogen. Therefore the precursor of that hormone must have been cholesterol. In turn, the precursor of cholesterol is acetic acid. Thus, the acetic acid in a salad dressing eaten by a pregnant woman may be incorporated into a hormone molecule by the virtuoso synthetic processes of the human body.

Isotopes are a wonderful tool of detection. With their aid we can follow the path in the body of a particular hydrogen atom in a particular molecule. Questions which seemed unanswerable in pre-isotope days are being answered routinely. Most of our knowledge of the intermediate metabolism—the overall fate from absorption to excretion—of fats and amino acids we owe to isotopes.

But isotopes won't yield miracles. When the newspapers and magazines became isotope conscious they began to

[5] The feeding of normal constituents of the diet containing stable, non-radioactive isotopes is perfectly harmless. In this case the labeled cholesterol was eaten by the biochemist's wife.

predict medical millennia just around the corner, produced by isotopes. The dragons of cancer, heart disease, or whatever ails one were being slain by the isotopic swords. While making the public research conscious is extremely valuable, it is cruel to raise false hopes. Isotopes are a tool, a good tool, but just one of many tools.

A tool by itself has never built anything. The scientists whose minds and hands wield the tools, are the architects of medical research. Only the ideas of men and women who can dream them will penetrate the startling complexity of a living cell. Only when we have a clear view of the normal pathways in the labyrinth of the cell will we be able to trace the monstrous blunders which lead into the cul-de-sacs of cancer or arteriosclerosis.

Bold, direct frontal assaults have recently been made on cancer, using radioactive isotopes as weapons. But we cannot as yet control the range of the weapon; it is shooting friend and foe alike. The sensitivity of cancer cells to high-energy radiation such as X rays and radioactivity is well known. However, normal cells are vulnerable too, and unfortunately the staying power of the normal cell under the impact is only very slightly greater than that of the wildly growing cancer cells. There is always some destruction of normal tissues as well.

The dream is to send packages of radioactivity by special delivery into only the diseased cells.

The immense energy of the uranium pile can produce a variety of highly radioactive elements, including phosphorus. This element, in the form of phosphates, is known to concentrate in rapidly growing cells. It was a bright hope that large doses of radioactivity could be delivered to cancer cells by means of radioactive phosphates. One tragic example will illustrate the difficulties we must overcome.

A patient with inflammatory cancer, a hopeless, wildly spreading type, was treated with a large dose of radioactive phosphate. His recovery was miraculous; his cancerous lumps receded; he returned to his normal work. We dared to permit a bright sliver of hope to puncture our gloom on cancer. Unfortunately the patient had a relapse; in a month or so the cancer flared up again. This time a larger dose of radioactive phosphate was administered. The cancer subsided once again but in a few days the patient showed symptoms of radiation sickness. He died, like many who had at first survived the blast over Hiroshima, from the destruction of normal tissues by radioactivity.

We are back to Ehrlich's old problem: to carry the "magic bullet" to specific cells. This time the bullet has, not arsenic, but radioactivity in its warhead; otherwise the problem has not changed. As the French say, "The more things change, the more they remain the same." We can but hope that there is another Ehrlich, not too far away, who will direct the bullet to its mark before the biological scientists unravel the labyrinth of the cell.

6 Amino Acids and Proteins

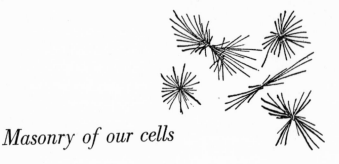

Masonry of our cells

"Let them eat cake," was the reputed dietary advice of Marie Antoinette to the undernourished poor of Paris.

Came the revolution, the poor, instead of more food, got more advice: "Let them eat glue." This dietary exhortation came from Cadet de Vaux, a physician, who urged the Parisians to make soup out of glue, or gelatin, guaranteeing it to be as nutritious as beef soup. The government issued official proclamations endorsing the new substitute for Marie's cake. The Institute of France and the French Academy of Medicine added their authority, praising the ersatz food and cajoling the starving Parisians to become converts to it.

But the Parisians would have nothing to do with such newfangled nonsense. A political revolution they took in their stride, but a revolution of the stomach, that is a serious matter.

About 150 years later biochemists proved the wisdom of the adamant Parisians. There were attempts to evaluate the food value of gelatin long before the biochemist brought some order to the chaos of the field of nutrition. The most interesting attempt in this prehistoric era of nutrition was made by M. Gannal, a manufacturer of glue who boldly resolved in 1832 to test the food value of his product. He noticed that the rats which infested his factory ate the raw materials—the tendons, cartilage, and skin of animals—but snubbed his product, the glue, which was obtained by cooking these animal wastes with water.

Gannal conjectured that perhaps rats were merely fastidious about the taste and odor of glue. He therefore decided to perform his feeding experiments on humans. He is to be commended for not urging the consumption of his product on the poor; he fed it to his wife and three children and he himself joined them in their dreary diet. They ate glue, and, to make it somewhat more palatable, glue and bread, for weeks. The result was disastrous. They had violent headaches and intense nausea, and when it became apparent that their health was rapidly deteriorating, M. Gannal reluctantly called off the experiment. He sadly concluded that his product has no food value, indeed it is harmful.

The only thing more difficult than the introduction of a new, fruitful idea, is the banishing of an old, fruitless one. The feeding of glue, or in its more purified form gelatin, kept cropping up in scientific and medical literature for the next hundred years. Convalescents in hospitals, nursing mothers, and infants were fed gelatin. Fortunately in most cases the feeding period was brief. Nor was gelatin the sole protein in the diet of the experimental humans. The conclusions drawn were varied, depending upon the susceptibility of the subjects and of the experimenters to autosuggestion.

The field remained chaotic until biochemists entered it with their zoos of experimental animals. First of all, it was established that proteins are an absolute essential in the diet of rats and of dogs. Caged animals, which could not forage for food, were kept on diets completely devoid of proteins. They rapidly lost weight and unless they were rescued with meals of proteins, they died.

The next question was: Is it the proteins themselves or their amino acids that are essential to the animals?

A wholesome protein, casein from milk, was cooked with acid until it was whittled down to its amino acids. Rats that had no protein in their diet were fed this mixture of amino acids. The rats thrived. This should not be surprising; after all, what happens to the proteins in the alimentary canal of animals? They are broken down by the enzymes into amino acids. Animals do not absorb whole proteins into their blood streams; they absorb only the amino acids. Whether the protein is crumbled down by acid in the flasks of the chemist or by enzymes in the stomach and intestine of the animal apparently makes no difference (except for the destruction of one of the amino acids by the hot acid).

The emaciated skeletons of men liberated from the Nazi and Japanese prison camps benefited from this knowledge. The shrunken internal organs of these people, after months of starvation, could not tolerate the food they craved. Their stomachs had forgotten from the months of abuse the know-how of protein digestion. They were saved and rebuilt by feeding them the amino acids from predigested proteins, which their abused internal organs could tolerate.

Progress in science is similar to a duel with the mythical multiheaded Hydra. For every question answered, other new questions crop up. Once the biochemist proved that it is the amino acids, not the whole protein, which is essential, he had to tilt with a brand new and far more difficult question. Are all amino acids of equal value to the animal? The obvious way to resolve this question is to feed all of the amino acids to experimental animals and then keep withdrawing the amino acids from such a diet, one at a time.

While this scheme of eliminating amino acids is simple and obvious it could not be carried out until recently because of technical difficulties. Individual amino acids can only with difficulty be cajoled out of a natural mixture of them. They are so alike in chemical behavior that they manage to hide each other from the chemist who is bent on extracting one of them.

Biochemists therefore turned to some natural proteins which were known to be lacking one or more amino acids. Gelatin was found to be such an incomplete protein. Three amino acids are completely missing from it and two others barely put in an appearance. Gelatin is by no means the only protein so poorly endowed with amino acids. Zein, a protein from corn, and gliadin, a protein from wheat, are both lacking in some amino acids.

Rats cannot live on a diet complete in every other way but containing as the only source of amino acids these impoverished proteins. Their growth is stunted, and unless help comes in time, they die. The help is either a wholesome protein like casein or the original, incomplete protein fortified with the missing amino acids.

The growth curves of rats on such a diet illustrate vividly the need for these amino acids.

Weaned, litter-mate rats are used in such experiments to rule out individual variation.

A rat on a complete diet containing casein as the protein grows like this:

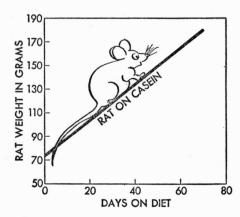

But if in the same nutritious rat meal, casein is replaced by zein, the protein from corn, the rat will lose weight at an alarming rate.

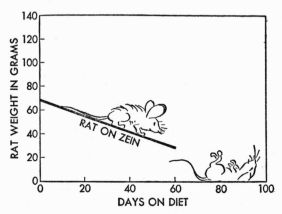

But if we come to the rescue of the moribund rat and throw him a life belt of the two missing amino acids, tryptophan and lysine, the rat will perk up and will begin to gain weight with sufficient speed to walk right off the graph.

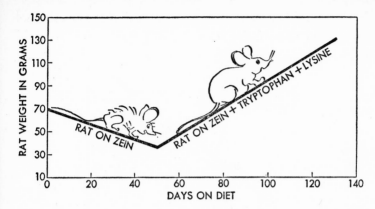

The next question in this logical sequence was: Does the rat find every single amino acid essential in its diet? Very little work was required to answer this question. Casein is one of the best proteins; all suckling animals thrive on it. Yet this wholesome protein lacks one amino acid. We must conclude, therefore, that all amino acids are not essential, only some of them are.

But which of the twenty amino acids are essential? Had there been twenty different proteins each lacking a different acid, our task would have been easy. Those proteins which could not support growth, we could conclude, lack an essential amino acid. Unfortunately, the various proteins found in nature were not designed for the convenience of the biochemist. There are no such proteins with prefabricated amino acid deficiencies. We had to sort out the essential and nonessential amino acids the hard way.

Biochemists have been nibbling at this problem from the start of the century without making too much headway. About twenty years ago William C. Rose, a biochemist at the University of Illinois, started a series of well-designed experiments which were to serve as a powerful beacon illuminating this whole complex field. We learned, under the light of this beacon, that there are no less than ten essential amino acids and that a protein is worth no more than the amount of these amino acids it contains. Dr. Rose decided to keep rats on a collection of pure amino acids rather than on proteins. In this way he could leave out one amino acid at a time and study the effect of the omission on his rats.

It is difficult to convey to the reader what an onerous task this was. Growth studies are long drawn out affairs. In some cases rats had to be continued on the same diet without a day's interruption for six to seven months. To test twenty different amino acids meant keeping twenty different groups of rats on twenty different diets.

The pedigreed rats [1] used by the biochemist are sensitive creatures; they react to the slightest changes in their environment. Noise, variations in temperature, changes in lighting, the freshness of their food, all affect their growth rate.

Only a few of the amino acids were to be had commercially, and those were prohibitive in cost. Since these experiments were conducted years before the atom bomb became the exclamation point to the scientist's plea that research pays, money for research was not easily forthcoming.

Rose and his graduate students at Illinois labored long at the accumulation of amino acids. Some they isolated from mixtures of amino acids obtained from proteins; others they made synthetically in the laboratory. How simple that sounds. But what a tediously difficult task it

[1] The genealogy of these rats is better known than that of the noblest entry in the Almanach de Gotha.

was even for the expert hands of scores of graduate students trained in that remarkable school of chemistry.

When all the known amino acids were assembled they were fed to rats which were receiving no proteins at all. The rats failed to grow. They could be induced to grow only if the amino acid mixture was supplemented with a bit of casein. Dr. Rose concluded that something else besides the twenty known amino acids must be present in casein and that this unknown substance is also essential for the rats. To track down the unknown factor he began one of those tedious, painstaking searches which the reader by now must recognize as an occupational hazard of the biochemist.

Casein was cooked with acid and the resulting amino acid mixture was put through a variety of chemical separations. Each of these fractions was fed to rats along with their diet of amino acids. One fraction when added to the diet of the known amino acids enabled the rats to grow. This fraction, after appropriate purifications, yielded a brand new amino acid of whose existence we had not even dreamed, so well had it been hiding among its brethren. Rose started once again to feed his rats all the previously known amino acids plus the new one. The rats thrived.

Then he started a group of rats on a diet containing all the amino acids but one, and he measured their growth. Another group of rats was put on a diet lacking another one of the amino acids and this was repeated until all of the twenty-one amino acids had their turn at being left out in this most elaborate of musical chairs games.

Ten different groups of rats failed to grow; ten amino acids are essential to the rat. They can do without the other eleven amino acids, but if just one of these ten essen-

tial amino acids is missing the rat behaves as if it were receiving no protein at all; it cannot grow.

The need for protein, then, is the need for these ten
essential amino acids. "By their essential amino acids shall
ye value them," should be the injunction to guide us in our
choice of proteins. For it has been found more recently,
that in the need for the essential amino acids, we humans
are akin to the rat. We are not quite as exacting as the rat
in our dietary requirement. Our cells can make one of the
amino acids and thus we require only nine of them as
absolute essentials in our diet.

Now we know why the Gannal family became ill on
their diet of gelatin: they lacked some essential amino
acids. However, no harm results from eating gelatin. The
writer would not recommend gelatin as the sole source of
protein in the diet, but as a low-caloried, decorative ad-
junct to a meal it is useful.

The distribution of the essential amino acids in foods is
unfortunate. We find that only the expensive animal proteins—meat, cheese, and eggs—are well endowed with
them. The cheaper plant proteins are either low or completely lacking in some of the essential amino acids. Some
bean proteins, especially those of the soya bean, approach,
but never reach, the animal proteins in their essential
amino-acid value.

The impoverished nature of vegetable proteins is interesting in view of the fairly widespread fad for vegetarianism. In the election of 1948 there was a candidate of the
Vegetarian Party for President. The constructive program
this party offered was the slaughter of all food animals,
and the conversion of all grazing lands into farms to raise
vegetables in sufficient abundance to make up for the loss

of meat. Whoever concocted this little scheme knew nothing about nutrition and even less about agriculture. The production of meat is by far the most efficient, indeed the only way to utilize over a billion acres of our marginal land. Close to a billion acres of ranch land are not fit for raising anything but grass. This land would be totally unproductive for human nutrition without grazing animals, which alone can convert the grass into meat. Of another half a billion acres of pasture land about three fourths are too hilly for plowing. (We must concede one virtue to the vegetarian platform: it is explicit, a rare quality in this very specialized kind of literary endeavor.)

The practice of absolute vegetarianism from early childhood would be disastrous. The only reason that the effects of the poverty of essential amino acids do not become apparent in vegetarians is simply that there are no absolute vegetarians. The possible exception is the Chinese coolie with his daily bowl of rice, and he is not noted for robust health. Most vegetarians eat milk, cheese, and eggs.

Even the most celebrated vegetarian, the late George Bernard Shaw, had strayed from the true path: he used to take liver pills. According to the late Alexander Woollcott, Shaw referred to these furtive adjuncts of his vegetarian diet as "those chemical pills."

Partisans of vegetarianism had triumphantly ascribed Mr. Shaw's zestful long life to his partially vegetarian habits. Advocates of health fads indulge all too frequently in this kind of semi-reasoning. Well-controlled experiments on the human diet, running for decades on the same individuals, are impossible to achieve. But hope for longevity is fervent, and it is so easy to draw dietary conclusions to nurture that hope.

Those bent on partisan explanations of longevity are particularly apt to commit the "after this, therefore because of this," error of logic. Everything a man has ever done preceded his old age. Sibelius has been a chain smoker of potent cigars most of his life. Should the composer's hale longevity be ascribed to his smoking habits? The amounts of essential amino acids needed are fortunately so low that a person on an average, well-balanced diet can dismiss all concern about them. For example, the British Ministry of Health advocates the daily consumption of not quite two ounces of a "first-class protein," meaning meat, fish, cheese, or eggs. American authorities advocate about two and a half ounces of mixed proteins per day for an adult. Obviously anyone who eats an egg or two and a fair serving of meat or fish a day is well provided with "first-class protein."

The writer emphasizes the adequacy of the average American diet in amino acids, to provide a storm cellar for the reader. He predicts that such a storm cellar will be badly needed in the next few years. The ad men are beginning to discover amino acids. They will wallow in the twenty-one bizarre names the biochemist has given to the amino acids. We will be scolded for not eating enough serine and threatened with dire consequences from the lack of threonine; our social inadequacies will be ascribed to a lack of isoleucine.

Actually, hypoproteinosis, the condition induced by lack of good proteins, is seen only among people on very impoverished diets. Its symptoms are somewhat indistinct; vitamin deficiencies and the cumulative effects of squalor are invariably associated with it.

A multitude of functions in our bodies make the amino acids so indispensable to us. In the first place, they are the bricks of which our tissues are built. Twelve of these twenty-one bricks can be made in our own bodies; that is why they can be left out of the diet with impunity. If one of these twelve amino acids, for example, alanine, is omitted from the diet, we can make enough of it for our needs.

One of the stages in the metabolism of sugars, is pyruvic acid. Alanine, the amino acid for whose absence we are about to compensate, and pyruvic acid are very similar in chemical structure. They both consist of a chain of three carbon atoms, two of which—the two end ones—have the same atoms attached to them. They differ only in the atoms that the middle carbons bear. In pyruvic acid there is an oxygen atom attached to the middle carbon, in alanine an atom of nitrogen.

If we happen to need some alanine to build into our body proteins and none of it is forthcoming from the diet, the enzymes in our liver improvise the alanine. They marshal a molecule of ammonia and a molecule of pyruvic acid and clip them together to form alanine. The oxygen atom, which is set free from the pyruvic acid, soon gathers up two hydrogens and they swim away as a water molecule.

Thus we can make an amino acid out of a by-product of sugar metabolism. This is a great asset. We are not slavishly dependent upon our diet for twelve of the amino acids. We can produce each of these by making over some fragment from the utilization of our sugars and fats, upholstering such a fragment with ammonia. But the nine essential amino acids that we cannot make must come to us ready made. We can do a bit of an assembly job, we can add am-

monia to the appropriate carbon skeleton, but we cannot
fabricate the carbon skeleton of the nine essential amino
acids. As the mason needs a specially designed keystone
when building an arch, so our enzymes need the nine prefabricated essential amino acids for building proteins.

Nor is the assembly into protein molecules the only role
of amino acids. They are the busiest molecules. They are
made into hormones and into body pigments, and they are
unleashed to disarm poisonous invading substances.

It is beyond the scope of this book to describe the special
role each amino acid performs in the body, but one, methionine, will be used as an example. Methionine was chosen
for two reasons: it has an interesting history and it performs
interesting functions in our cells.

Methionine makes up 3 percent of casein and it is essential to rats, men, and microorganisms. Yet we had no idea
of its existence until 1922, or of its structure until 1928. It
is another one of those biologically important substances
which was discovered by work not on animals but on microorganisms. It was discovered by one of the greatest of contemporary bacteriologists, John Howard Mueller, whose
chief interest was the dietary requirement of diphtheria
bacilli.

This aspect of bacteriology, the nutritional need of microorganisms, was in chaotic disarray at the time. Every science passes through such a muddled period, until coordi-
nating principles are found to weave some pattern out of
the mass of apparently unrelated bits of information gathered by the pioneers of the science. Chemistry did not
emerge from alchemy until the end of the eighteenth century; bacteriology, a much younger science, started to
emerge from its chaotic morass in the second decade of this

century. Bacteriologists had tried everything at frantic random to grow bacteria away from living host animals. The prescriptions for raising bacteria, until recently, read like a cookbook for apprentice witches. A concoction might be made of a pig-heart infusion, with a dash of yeast extract, and a soupçon of beef blood. Today, many microorganisms can be grown on completely synthetic diets. As for the others, just give us time, they will eat out of our hands yet—synthetic food.

Mueller was interested in the amino acids needed by diphtheria bacilli for growth. Instead of the bacteriological witches' brew he raised them on acid-cooked casein. He then put the amino-acid mixture through the usual chemical fractionations and tested each fraction as a diet for the bacilli.[2]

From one of these fractions he isolated a hitherto unknown amino acid, methionine, which proved to be essential for the diphtheria bacilli. About ten years later Rose found the same amino acid essential for the rat and only a few years after that the amino acid was being used in the therapy of some diseases of the human liver. The history of methionine offers but one more example of the unforeseen bounties that can accrue from basic research. That not only the public, but even pharmaceutical houses were blind to the value of research is all too clear from Mueller's experience. Recently he wrote: "The writer recalls somewhat grimly the difficulties encountered in 1920 while attempting to enlist cooperation [of pharmaceutical companies] in getting a hundred pounds of casein hydrolyzed with sulfuric acid, from which methionine was eventually

[2] This is of course the very process Rose was to repeat about ten years later in his search for the essential amino acids of the rat.

isolated." The current profit from a day's sale of methionine would more than cover the cost of the little favor for which Mueller was pleading.

What does methionine do that makes it indispensable to man, beasts, and bacilli? Methionine is one of the two commonly occurring [3] amino acids which contain, in addition to the usual elements, the element sulfur. The fate of the sulfur of methionine in animals has been charted with the aid of isotopes. Methionine has been made in which the sulfur atom is not an ordinary sulfur but a radioactive isotope. This sulfur can be traced by the radioactive messages it sends to a receptive Geiger counter. It turns up in the other sulfur-containing amino acid, cystine, which makes up almost 15 percent of the proteins of the skin and hair. These are very special proteins: they are tough and they are insoluble in water. We should be especially grateful to some creeping ancestor of ours who first acquired such a skin by a fortuitous mutation. Imagine having a protein such as egg white for the skin. Getting caught in a rain would be fatal; our tissues would trickle away and nothing would be left but our bones, in the midst of a puddle of tissues.

One of the roles of methionine, then, is to provide the raw material sulfur, for one of the amino acids of the skin (and of other tissues as well).

Attached to the sulfur atom in methionine is a very simple group of atoms—one carbon, loaded with three hy-

[3] There are more than twenty-one naturally occurring amino acids. Quite a few others are found in more rare sources; for example, the octopus yields octopine. While these rare amino acids apparently play no role in mammalian nutrition, we treat them with respect. One such amino acid was found by a Japanese investigator in watermelon seeds. A few years later this amino acid was found to be not so rare after all: it plays a stellar role in the formation of urea in the human liver.

drogen atoms—called the methyl group. It is the simplest pattern in organic chemistry. The methyl group of methionine, too, has been traced with the aid of isotopes. After methionine is eaten, the labeled methyl groups from it turn up in several different substances in the body, substances of vital importance: for example, in choline, a compound with many roles. Choline aids in the metabolism of fats; it is also part of the impulse-sending mechanism in nerve tissue. It contains three methyl groups in its architecture. These methyls are supplied by methionine.

A deficiency of methionine and of choline can be disastrous for the animal. Its liver becomes diseased and degenerated. Choline and methionine have become valued tools in the treatment of certain diseases in the past few years. From the curiosity of a bacteriologist about the dietary needs of the diphtheria bacilli, a medication for humans was harvested!

That the methyl group must come prefabricated in the diet of an animal was a great surprise. Animals must depend on plants to make this simplest of fragments for them. This is an especially puzzling riddle when we recall that animals can make complex things like fatty acids of twenty-four-carbon-chain length.

(Some of these statements about the inability of the organism to fashion its own methyl groups have become obsolete since this chapter was originally written. It has been found recently that in the presence of ample supplies of vitamin B_{12}—about which, more in the next chapter—animals can make methyl groups. However, the writer left in the obsolete, or rather, false information, purposely. He wished to show how rapidly our views must change as new information is gathered in this very active field.)

And what happens to the remnants of the methionine molecule? After detaching the methyl group and the sulfur atom, we are left with four carbons to which is attached a nitrogen atom. All we know about this fragment is that its nitrogen can be severed from the rest of the molecule to form ammonia which, in turn, can be incorporated into urea and then excreted. What happens to the rest of the molecule is still unknown. It may be just metabolized to carbon dioxide and water; or it may be used to make other substances such as fatty acids. Then again it may have some unknown specialized role. A breakdown in its unknown metabolism may be the cause of any one of our metabolic diseases.

Finally, methionine as a whole enters into a combination with other amino acids to form that most marvelous of substances: the protein of our tissues.

The protein molecule is nature's masterpiece of complexity. In the elaborate pattern of that molecule is locked the secret of life. The proteins are the mechanics of life: they fabricate its tissues, regulate its energy, and assure its perpetuation.

There is no extensive storage of proteins in our bodies. We can store huge amounts of fats in our fatty depots and we store sugar in the glycogen depots, but we store no proteins in excess of the amounts needed to make up our tissues. On the one hand this is a serious drawback, since we must, therefore, depend on a steady supply of essential amino acids from our diet. But then, many proteins are active enzymes; indeed some of us think that every protein molecule is an enzyme. The unchecked accumulation of highly potent protein molecules could become disastrous: we might grow to monstrous size and an uneven distribution of proteins in our bodies could wreak havoc with the delicately adjusted balance of the various parts of the whole individual. The accumulation of large blobs of fat around our abdomen does damage only to our vanity. The deposition of similar amounts of protein would be lethal. Even hibernating animals, which must lay in a large store of food in their bodies for a long siege of starvation, store only fats, not proteins.

If an animal happens to receive an overgenerous supply of amino acids, in excess of its immediate needs, it metabolizes them into urea, carbon dioxide, and water; or, if it requires no immediate source of energy, it can convert the excess amino acids into sugars and fats, the depots of which can be readily augmented. Earlier in this chapter the process by which an intermediate in sugar metabolism, pyruvic acid, can be converted into the amino acid alanine, was outlined. If, on the other hand, an animal is encumbered with an excess of alanine, it reverses this process: it converts the alanine into pyruvic acid. Certain enzymes— mostly in the liver and kidney—remove the ammonia from the amino acid and replace it with an atom of oxygen. Other enzymes combine the newly formed ammonia with carbon dioxide to form urea which is then excreted. Thus two fairly toxic waste products, ammonia and carbon dioxide, are efficiently eliminated in one step. The molecule of pyruvic acid is either further metabolized to carbon dioxide and water or two molecules of it are fused together to form a molecule of glucose. The pyruvic acid can, by a more elaborate process, be converted into a fat, too. Thus the proteins in our diet are ready sources of carbohydrate and fat. It is well to remember this, since the nonfattening nature of proteins has been overemphasized by the designers of reducing diets. While it is true that proteins have lower caloric value per unit weight than fats or carbohydrates, nevertheless, unfortunately for those bent on acquiring a more stylish figure, a thousand calories coming from lamb chops are just as fattening as a thousand calories from rice pudding.

There *is* a condition in which there is excessive protein synthesis: in cancer protein synthesis runs riot. Masses of protein accumulate until internal organs are strangled and, eventually, life itself is choked off. Since we know very little about the intimate structure of the protein molecule and less about the mechanism of its formation, we consequently know very little about cancer. And what we do not understand we cannot control.

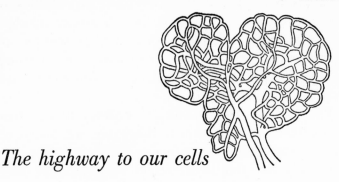

The highway to our cells

"BLOOD IS A TRULY REMARKABLE JUICE," said Mephistopheles to Dr. Faust as the two went through the sealing of their contract.

Mephistopheles showed rare biochemical insight: blood *is* a truly remarkable juice. It is a juice to which we owe much. We owe to it our size, we owe to it our brain, we owe to it our wonderfully complex physiological existence.

It was the great French physiologist, Claude Bernard, who pointed out, in 1878, that the evolution of the highest forms of life has been made possible by the liquid *milieu intérieur*. "The living organism," he wrote, "does not really exist in the *milieu extérieur* [the atmosphere for terrestrial animals; salt, or fresh water for those who had not invented lungs] but in the liquid *milieu intérieur* formed by the circulating organic liquid which surrounds and bathes all the tissue elements."

Complex life is possible for the biological organism, only with adequate means of transportation from organ to organ, just as complex social life is made possible only by transportation facilities from community to community.

Blood is the highway to the cells of our tissues. Without it the cells, even those on the surface of our skin, would perish. They would lack oxygen; they would lack food; they would be killed by poisons of their own making.

The center of this most wonderful system of transportation is, of course, the heart. It pumps, in a lifetime of seventy years, about two billion times, and it pushes on its path a hundred million gallons of blood. From the right heart to the lungs, from the lungs to the left heart, from there, through the arteries into the capillaries of the tissues, back through the veins into the right heart, 'round and 'round goes the blood in its wondrous, uninterrupted circle, performing many chores on its rounds. It takes carbon dioxide from the cells and exchanges it for oxygen in the lungs. It is a traveling department store of foodstuffs. It carries everything a cell needs: amino acids, fats, sugars, vitamins, and salts. A hungry cell in one of the outlying districts, say in the toe, extracts from the blood swishing by whatever it requires: a few million molecules of glucose, a hundred thousand molecules of vitamin B_6, a few thousand cobalt atoms. Each cell, however far removed, is thus as well provisioned as a cell in the heart itself.

In addition to foods, the blood carries a variety of other wares: hormones to stimulate laggard cellular mechanisms; antibodies to battle invading poisons; clotting agents to seal breaches in its cyclic path. Furthermore, the blood distributes the heat evolved from the furnace of the cell, thus maintaining a uniform temperature throughout the body.

Finally, the acidity of tissues is kept within tolerable bounds by the blood. Life is fenced in within very narrow limits of acidity. A variety of acids are produced by the metabolic activities within the body cells. Carbon dioxide, lactic acid, and uric acid all tend to acidify the human body. Excess acidity slows down many enzymes, and unless the acidity is counteracted, life itself slows down and eventually halts. Many bacteria—for example, those which turn milk sour—are destroyed by their own metabolism. They produce lactic acid and throw it out into the surrounding fluid. Soon, so much acid accumulates that they become the victims of their own sewage. The blood contains powerful neutralizing mechanisms ready to pounce on the acids cast off by the cells.

How does this "remarkable juice" perform its many functions? It consists of both cells and a variety of noncellular, dissolved materials. About 45 percent of the blood is composed of the red cells. They are tiny red discs (five million of them are packed into a volume the size of a sugar crystal) containing the red protein hemoglobin. This is the truck on which oxygen and carbon dioxide are shuttled back and forth. It is an unusual truck, the hemoglobin molecule, it can carry only one five-hundredth of its weight of oxygen.

A hemoglobin molecule is made of a colorless protein, globin, and an iron-containing pigment, heme. Oxygen unites with this complex molecule in the lungs forming a definite chemical combination which, however, is easily decomposed in the capillaries, liberating oxygen to the gasping tissue cells. Once freed of the oxygen, the hemoglobin forms a temporary chemical alliance with carbon dioxide which is then ferried to the lungs.

Unfortunately, hemoglobin is not very discriminating as it forms its chemical alliances. Life can be snuffed out by carbon monoxide poisoning because of this lack of discrimination by the hemoglobin. It combines with carbon monoxide gas with far greater avidity (two hundred times greater) than with oxygen. Thus, if one inhales a mixture of oxygen and carbon monoxide, the two gasses compete for the favors of the hemoglobin, and oxygen has odds of two hundred to one against it in the contest.

Death from carbon monoxide asphyxiation is caused simply by the lack of free hemoglobin to transport oxygen to the starving tissue cells. The hemoglobin-carbon monoxide combination is actually not poisonous. City dwellers invariably have about one percent of their hemoglobin tied down to carbon monoxide, and tobacco smoking immobilizes as much as five percent of the hemoglobin. (Carbon monoxide is produced by the incomplete burning of the tobacco.) Fortunately, the combination between hemoglobin and carbon monoxide is not permanent. The gas is quickly swept out of the system, unless, of course, enough was absorbed to overwhelm the victim.

Red cells are highly specialized for their role as trucks for the transport of oxygen and carbon dioxide. Their metabolism is very low; they are stripped down to such an extent that the mature cells lack even a cell nucleus. They are nourished by the metabolism of other cells as long as they are useful, but when in old age they falter at their tasks they are liquidated. There are special cells in the blood vessels of the spleen and liver which unceremoniously devour the aging red cells. These cannibalistic phagocytes (eating cells) are constantly trapping and dismembering the more sluggishly moving red cells. The iron is salvaged from the wreckage but the pigment, heme, is piped into the gall bladder from where it is discharged into the intestine. What happens to the protein, the globin, we do not know.

Why the red cells must die is but one of the unsolved mysteries connected with them. There are many others. How do the phagocytes select their aging prey for the slaughter? Our instruments cannot distinguish between the young and old red cells. Are there stigmata of age recognizable, so far, only to the phagocytes, or do they merely fall upon the laggard cells?

All cells, young and old, travel under the impetus of the pumping heart. What makes an aging cell slow down in the capillaries? Is there a definite retirement age for red cells or are they destroyed at random?

Only to the last question do we have an unequivocal answer. The meteoric existence of the red cells was known for a long time but estimates of their career varied from five days to two hundred days. Recently the life span of the red cell was clocked with isotopes. The counting of the days of the red cell was an unexpected by-product of an entirely different project—as so often happens in scientific research.[1]

The metabolism of the simplest of amino acids, glycine, was to be studied in humans. A biochemist at Columbia University made two ounces of glycine which contained not the ordinary isotope of nitrogen but its rare, heavy isotope, and, showing the ultimate in confidence in the purity of his preparation, he promptly ate it. For weeks thereafter he obtained samples of his own blood and measured the

[1] A biochemist who completes merely his original projects is rather limited. The chemical ways of the cell are so much more complex than we can at present imagine, that, during the course of almost any project, mechanisms of far greater interest than the originally visualized one are invariably exposed to observant eyes. This is why freedom for the investigator to follow up unexpected, chance findings is so essential.

heavy nitrogen isotope in the various fractions of it. The largest concentration of the isotope was in the red pigment of the hemoglobin. Apparently this pigment, heme, is fashioned out of the glycine molecule, hence the high concentration of the heavy isotope in it.

This is one more example of the great synthesizing ability of animals. Glycine, which contains but two carbon atoms, can be marshaled by the tissues of the bone marrow, the birthplace of the red cells, into the elaborate structure of heme which contains 34 carbon atoms.

This unexpected discovery pointed a way to the investigator for the measuring of the life span of the red cell. Only those red cells which were made on the day or two following the meal of labeled glycine would contain the heavy isotope of nitrogen. As these cells would be devoured at the end of their careers, the isotope-containing heme from them would be voided through the bile and lost from the body forever. Therefore the disappearance of the heavy isotope from the heme would mark the death and disposal of the cells produced on the day of the isotopic meal.

Small samples of blood were tapped almost daily. (This biochemist should be doubly grateful to Pregl and the microchemistry he founded. His work was speeded and his blood spared by working with very small amounts of it.) The amount of heavy nitrogen remained at a constant level for about eighty days. After that it began to disappear. From a mathematical analysis of the complete data we can calculate that the average life span of the human red cell is one hundred and twenty days. The "average" is emphasized, for there was a small amount of heavy nitrogen isotope left even after one hundred and thirty days. Apparently, red cells, just like the whole organisms of which they are a peripatetic part, vary in their life span. Some red cells exceed, and some fall short of the six score days.

The rate of destruction and of synthesis of red cells is prodigious: about ten million red cells are born and about the same number die every second in each of us. The number of red cells can increase for a variety of reasons. One of the most interesting of these is prolonged stay at high altitudes. Since the concentration of oxygen at high altitudes is low, the body adapts itself to the emergency by making more cells and hemoglobin for the transport of the gas in short supply.

The ability to adapt to lowered oxygen concentrations is a valuable asset. For the greater the adaptability to changing environment, the greater are the chances of survival of an organism. The red blood corpuscles are not the only cells which can be augmented to compensate for an environmental or structural deficiency. If one kidney is removed from an animal the other kidney will increase in size, thus enlarging its functional surface and enabling it to carry the larger burden which falls upon it. The mechanism of the induction of the formation of new cells by a chemical stress—a decrease in the available oxygen or an increase in the amount of waste products awaiting disposal—is a challenging problem for the biochemist. For here he is coming to grips with the basic unsolved problem: the method of synthesis of a new cell, complete with its structural components, its enzymes, and its urgency to live. We are just beginning to explore the mechanism of adaptation with the tools of biochemistry. It was found, as in so many other cases, that the most fruitful of such studies are those made not with animals but with microorganisms. Certain microorganisms have the ability to cope with a nutritional deficiency by synthesizing apparently new enzymes which can enable them to use otherwise "undigestible" foods. The study of such adaptations is easier than the study of the adaptive formation of new cells in animals. In the case of microorganisms there is a synthesis of a new enzyme only, whereas the adaptive formation of new cells in animals is not much different from the synthesis of new cells under normal conditions. The biological scientist must be a prudent explorer of life's machinery. A thorough study of a small circumscribed area often yields rich rewards of knowledge, while forays, more ambitious in scope and area, have all too often been fruitless.

The foremost team of scientists studying adaptive processes in microorganisms is the one headed by Dr. Jacques Monod of the Pasteur Institute in Paris. They have been studying a microorganism which has the ability to produce a new enzyme for the utilization of sugar and thus escape starvation. The microorganism *Escherichia coli* cannot normally use milk sugar as a source of food. Milk sugar is a complex sugar made of two smaller sugar molecules, glucose and galactose. *E. coli* thrives on either of these simple sugars but it will not grow during short periods of incubation if its diet contains only milk sugar. It lacks the enzyme needed for the cleavage of milk sugar into its components. However, after prolonged incubation the life-saving enzyme begins to accumulate within the starving cells; the organism begins to grow. Such "adapted" cells can thereafter always use milk sugar, provided the diet is not interrupted. If there is an interruption by the feeding of the simpler sugars, the milk-sugar-cleaving enzyme disappears. Monod's group have shown that the adaptive enzyme is not produced completely *de novo;* a phantom precursor of it is present in the normal, unadapted cells. This explains why all microorganisms cannot form such adaptive enzymes: the ability and the structural precursors must be latent within the cell. Another fact established by the ingeniously designed experiments of this group is that milk sugar is not specific in enticing the cell to produce its new enzyme. Other substances, too, can provoke enzyme formation provided a key pattern of a few atoms present in milk sugar is also present in the substance which duplicates the action of milk sugar. Thus, we have learned that the formation of an enzyme is induced by a pattern formed by but a few atoms.

If the red cells are abnormally low in number or are deficient in hemoglobin, one is said to be anemic. Excessive hemorrhage is the simplest and most easily remedied cause of anemia. Fully one fourth of an animal's blood can be lost with impunity: the loss is made up in two to three weeks. This is the reason for the ease with which we can donate a pint of our blood, which is but one tenth of our total wealth of it. Another cause of anemia, particularly among infants, is lack of iron in the diet. Still other causes are poisons which destroy the bone marrow. Finally, the anemia which at one time was as dreaded as cancer is today is pernicious anemia. A little over twenty years ago this disease was as relentlessly fatal as an advanced case of inoperable cancer is today. But the disease is now mastered. Its conquest is a monument to the joint efforts of medicine and chemistry.

Until 1926 we had no idea of the cause of the disease. Infection, poisons, and cancer were all accused as the possible culprits. The methods of treatment were as varied as they were futile. There was an odd brake on progress against the disease: no experimental animals could be induced to come down with it. We can make dogs diabetic by performing a simple operation; we can transplant tumors; any of the infectious diseases can be implanted into almost any animal; but to pernicious anemia all but the human animal seemed to be immune.

Dr. George H. Whipple and his associates at the University of Rochester decided to study experimentally produced anemia even though it did not resemble pernicious anemia. Dogs were bled copiously and frequently to induce this simple anemia. The goal of the study was to see if we could intercede by some means and accelerate the rate of regeneration of red cells. Various dietary aids were tried and it was found that feeding beef liver to anemic dogs, helped their recovery.

After this discovery had been made, the next obvious step was taken by two physicians at Harvard, George R. Minot and William P. Murphy. They fed to their patients suffering from pernicious anemia huge amounts of liver and noted marked improvement within ten days.

As long as the patients were kept on a diet of about a pound of liver a day their improvement continued. Apparently there was something in liver which could protect a patient against the ravages of pernicious anemia.

That is as far as the physicians could go in the search for the cure. At this stage the biochemists took over and began to track down with their specialized searching tools the active principle in the liver. The approach of the chemist to such a problem is, by now, familiar to the reader. Using a variety of chemical manipulations—extraction, precipitation, evaporation—the chemist weeds out the inactive contaminants, testing at each stage the efficiency of his gardening by an assay of the potency of his preparations. Needless to say, the product is expected to become more and more active: a smaller and smaller weight of it should contain most of the original biological activity. Crystallized enzymes, vitamins, hormones, and essential amino acids are the testimony to the effectiveness of these methods of the chemist.

This search, however, was hindered by a particularly difficult obstacle. The only test for the presence of the active principle in the liver was the improvement it brought to human patients suffering from pernicious anemia and, with the easily administered liver therapy, the untreated disease was becoming more and more rare. One leading research center offered free hospitalization and medical care to any patient suffering from the disease in return for permission to standardize the preparations with which the patient was being cured.

Only twenty years after the search began was this particular obstacle removed. During those twenty years chemists managed, despite the lack of convenient testing, to concentrate the active principle several thousandfold. Whereas a pound of liver had to be fed to a patient each day, only one milligram (450,000 milligrams make a pound) of the concentrated material was needed per day when it was given in the form of an injection.

The outstanding biochemist engaged in the purification of the liver factor was Dr. Henry D. Dakin, one of the stalwart pioneers of American biochemistry. (The antiseptic solution which he devised, and which bears his name, was the life-saving antiseptic of the First World War.)

A point was reached in the purification of the liver factor

beyond which further progress seemed almost impossible. The product, even though enormously concentrated, was still far from pure. Testing the material became well nigh impossible: response by the rare human patients to the various concentrated preparations was almost uniform.

Then, in 1946, a brand new kind of dietary deficiency was reported in the rat. If rats were kept on a diet in which the protein source was alcohol-extracted casein, they failed to grow. (Of course the diet was complete in all the known vitamins.) The missing factor could be found in a variety of foods. It was also present in the commercial liver extract which was being used to cure human pernicious anemia. Let us call this unknown material the "rat-growth factor."

There had been studies on the nutrition of chickens, dating back twenty years, which showed that unless laying hens were fed some protein of animal origin, their eggs did not hatch normally. The factor present in animal proteins and essential for normal hatching was called the "animal-protein factor."

There were then three different unknown dietary factors: the pernicious anemia factor, the rat-growth factor, and the animal-protein factor. These three apparently different problems were tied together and the three factors were proved to be the same, as soon as a pure crystalline material was isolated. Before that could be accomplished, however, a rapid, consistent, and specific assay for these factors was needed. The assay for one of the three apparently divergent factors was provided by Dr. Mary Shorb. She found a microorganism—*Lactobacillus lactis* Dorner, or LLD for short—which requires for *its* growth the "rat-growth factor."

The use of microorganisms for assay purposes is a recent

development of immense value. We have seen earlier that yeasts and animals need the same vitamins, or that the amino acid methionine is essential in the diet not only of men but of mice and of microorganisms as well. Among the thousands of different microorganisms we can find some which require for their growth any of the dietary essentials of animals. Using them instead of animals for assay purposes is a great advantage.

In order to deplete their reserve of a dietary essential, animals sometimes have to be kept on a deficient diet for months. Indeed a deficiency often does not become apparent until the second generation. It takes months to raise a generation of rats but only minutes to raise a generation of microorganisms, so rapidly do they reproduce. Therefore, assays with microorganisms are a matter of hours.

The principle of such an assay is simple. The growth of bacteria is proportional to the amount of balanced diet available to them. If their diet contains all the essentials but one, that one factor, whether it is a vitamin, or an amino acid or a salt, becomes the limiting factor in the growth of the organism. Growth becomes proportional to the amount of the limiting factor available. For example, if a million bacteria grow on one microgram (1/28,000,000 of an ounce) of vitamin B_x, two million will grow on two micrograms of that vitamin. If now, on an unknown amount of Vitamin B_x 1.5 million bacteria will grow, we can then conclude, that the unknown sample contains 1.5 micrograms of the vitamin.

Such, in essence, is a microbiological assay. The assay of one of the amino acids will illustrate the value of such methods. Ten years ago the only reliable method for the determination of glutamic acid was a chemical one. It took a year to run one such assay. Today a skilled assistant can complete a score of determinations in two days, using microorganisms as aides in the task.

Dr. Shorb found a microorganism—LLD—which needed something for growth besides the then known dietary factors. Her LLD factor was found in the same foods which contained the rat-growth factor. Liver extracts were the best source of the LLD factor. Dr. Shorb suggested that perhaps the LLD factor and the pernicious anemia factor were the same. If this were so, then here, at long last, would be an assay for the pernicious anemia factor other than a test on a human patient.

Within a year chemists at Merck and Company isolated the pernicious anemia factor in pure crystalline form. It was the LLD factor, the rat-growth factor, and the animal-protein factor as well. Since the product was now a definite, pure compound and not just a vague "factor" it was entitled to a new name. Vitamin B_{12} was the name given to the shiny red needles which cure pernicious anemia, enable LLD to grow, permit normal hatching of hens' eggs, and allow rats to grow normally.

Vitamin B_{12} is the most potent of any of the known biologically active materials. It is effective in even smaller amounts than is biotin. The late Dr. Randolph West of Columbia, one of the foremost clinical experts on pernicious anemia, who was Dr. Dakin's clinical collaborator in the early days of the search for the anti-pernicious anemia factor, was, very appropriately, the first to report on the clinical potency of Vitamin B_{12}. He found that as little as 3 micrograms, or one ten-millionth of an ounce, when injected into a human, started an immediate improvement in the patient's blood picture.

At this writing the structure of Vitamin B_{12} is not known, but one remarkable constituent of it is known: it is the metal cobalt. That traces of cobalt are essential in the diet of the mammal was previously known. There are a number of such trace elements essential for health. Indeed, the list of elements essential in our nutrition begins to read almost like the chemist's Periodic Table of the elements. Now the specific role of one of them, cobalt, became clear. It is part of Vitamin B_{12} which is a coenzyme assisting several enzymes, among them those which fashion methyl groups (see page 119).

While knowledge of the functions of the inorganic salts is accumulating but slowly, it is known quite well what elements are present in blood and in what quantities. (It is easier, by far, to assay for an element than to pin down its biological function.) Blood contains the same salts as sea water and the salts are present in approximately the same ratio. However, blood is fivefold more dilute. In other words, the concentration of each element present in blood is one fifth that in sea water.

That their blood is but diluted sea water—with respect to inorganic salts—is telling circumstantial evidence for the marine origin of animals. The odds are enormously against the chance repetition of the same ratio of the same elements in the sea and in blood. The dilution of the blood of animals has been explained through studies made not by biologists but by oceanographers. Apparently, it is not blood which has become diluted but rather it is the ocean which has become more concentrated since animals arose from that vast aqueous cradle. It is known that the inorganic salt content of the ocean keeps increasing steadily. There have been some interesting extrapolations from the rate of that increase to estimate the era when the ocean had one fifth of its present concentration. It is said that the calculated time coincides with the time animals are estimated, from other lines of evidence, to have evolved from their marine precursors.

In addition to the inorganic salts, blood contains dissolved gases. Of these nitrogen is the most abundant and the least useful. (Oxygen and carbon dioxide are not merely dissolved; they are held in chemical bondage.) Nitrogen seems to have no function other than that of a diluent for oxygen. In the gaseous elementary form nitrogen is completely unusable and can be replaced by another gas, for example helium, in the gaseous mixture that an animal breathes. Indeed, that is precisely what is done with the gases pumped down to deep-sea divers in order to minimize the possibility of their experiencing caisson disease, or the bends. This is a neat application of some simple laws of physics to eliminate a dangerous occupational hazard. Caisson disease is caused by the greater solubility of nitrogen in blood at higher pressures. If the pressure on a diver is quickly released—by too rapid surfacing—the nitrogen dissolved during his stay in the depths is suddenly released, forming bubbles in his blood vessels. At that moment the diver's blood simulates in appearance the contents of a bottle of carbonated cherry beverage from which the sealing cap is suddenly removed. The release of gas bubbles in the veins and arteries causes violent pain, convulsions, and, in severe cases, death. Helium is quite as inert as nitrogen in our bodies but is far less soluble. Thus when inhaled even at great pressures it does not endanger the diver, for too little of the gas dissolves to liberate bubbles at atmospheric pressure.

The freedom of the whale from the bends—this occupational hazard of other deep-sea divers—has puzzled biologists for several generations. According to whalers a wounded whale can take a half a mile of line to the depths and can then surface with startling speed. An equally rapid ascent from two hundred to three hundred feet would surely release enough bubbles of nitrogen to kill a man. How does the whale tolerate the tremendously greater changes in pressure? It has been found recently that the whale harbors in its blood stream myriads of a certain species of microorganism which sop up its inhaled nitrogen. This microorganism, like those in the nodules of the clover leaf, is nitrogen-fixing; it has the ability to convert gaseous, molecular nitrogen into soluble compounds of it. As fast as the whale absorbs nitrogen from its lungs, the gas is sponged up by the nitrogen-combining enzymes of the tiny inhabitants of its blood and the whale can dive and fish at otherwise unapproachable depths.

The noncellular part of the blood, the plasma, contains 7 percent dissolved proteins. This, in addition to the 14 percent of hemoglobin, brings the total protein content of blood to 21 percent. That the hemoglobin is not freely dissolved but is packaged in the red cells is one more example of the remarkable chemical foresight of nature. A 21 percent protein solution would have tremendous viscosity and flow would be impossible. Blood would indeed be "thicker than water." As it is, the hemoglobin is safely wrapped up in the red cells and does not impede the easy flow of blood.

The free plasma proteins contain the clotting mechanisms and the antibodies. Great strides have been made, particularly since the start of the Second World War, in the separation of plasma proteins into their various fractions. The products have widespread clinical uses. Absorbable surgical threads and fibrous packing which stop hemorrhage during surgery are everyday tools of the surgeon, made from human plasma proteins.

So far, but a tiny fragment of our chemical knowledge of that "truly remarkable juice," blood, has been related. But this is only a conducted tour of the kingdom of biochemistry. It must not be assumed that the reader cares to take up residence there.

8 Cell Defense

Many are the foes which attack the cell. The assaulting hordes come in strange shapes and varied sizes and aim a diverse battery of weapons at their target. The cell fights back with sustained valor. If its armed vigilance falters, disaster befalls the cell fortress. The struggle starts at birth and continues relentlessly. There is no quarter, no armistice; only survival or death. That cells do survive is a miracle wrought by their defensive weapons—weapons of great variety and of astonishingly ingenious design.

What particular cell-guard weapons from the well-stocked arsenal are mobilized at any one time depends upon the nature of the invader. The cell can burn up the invading enemy; it can, by means of enzymes, alter the marauder to remove its sting; it can fashion special shock troops which, using their strands of protein molecules grapple with the invader until it is immobilized.

Let us observe some of the embattled units in this constant warfare.

The best defense against any foe is to prevent its penetration. The body's first line of defense is a tough skin, which, though quite effective, has, unfortunately, some weak spots. Bacteria lodge in the pores of the sweat glands and in the hair pits, causing pimples and boils. Poisons and bacteria can pour through the larger openings, the mouth, nose, and eyes. The stomach, however, is booby trapped against the bacteria: the high concentration of acid in its juices kills most of them.

If a simple, chemical poison enters the body it is handled with effective vigor. The defensive campaign follows a well-defined strategy. The largest route of entry for such poisons is, of course, the mouth. The first maxim of the strategy is: Absorb as little of it as possible. A poison can do us no harm while it is in the alimentary canal. The transient contents of that long tube are actually not part of the body; they can do us good or ill only when they gain admission into the blood or other tissues.

There is a remarkable screening performed during absorption. Before a substance can gain admission into the tissues it must pass the discriminating scrutiny of the cells lining the alimentary canal, and these are remarkably adept at excluding undesirables. For example, humans absorb a component of egg yolk, cholesterol, with ease. However, there is an astonishingly similar substance made by plants, sitosterol, which is not absorbed at all.

If a poison passes the selective barrier of the alimentary canal, the appropriate order of the body's high command is: Excrete it! Use the kidney, use the lungs or sweat glands, but excrete it. This is, of course, a selectivity too, but in reverse. Now the foreign substances are preferentially expelled.

If excretion fails the command is: Burn it up! To study

the fate of toxic substances in animals we inject them. In this way we by-pass the forbidding scrutiny of the alimentary canal. If we insert into the muscles of an animal some benzene, it is promptly oxidized (mostly to carbon dioxide and water), by the enzymes of the victim, and the animal is rid of the poison.

The enzyme systems which burn up toxic substances are not teleologically designed for this single, self-defensive purpose. These enzymes are always present and are usually performing their normal metabolic oxidations. But if a toxic, foreign substance comes along and happens to fit into the working pattern of an enzyme system, the animal benefits from the versatility of its enzymes.

It would be difficult to visualize how an organism could be equipped to handle any poison it might encounter with enzymes tailored a priori to that purpose. There was a period in the development of biochemistry when hundreds of different organic compounds were administered to animals in order to study their fate in the body. Many of these substances had undoubtedly never existed in the universe until zealous organic chemists strung them together. But still, even though the animal had never, in its whole evolutionary history, encountered these substances, it promptly oxidized them or made extensive alterations in their structure.

During the Second World War there was information that the Germans were manufacturing on a large scale a substance called diisopropylfluoro-phosphate. (Even chemists call it DFP.) Since DFP paralyzes its victims, it was feared that the Nazis might use it as a "nerve gas." DFP was therefore extensively studied by the medical unit of the Chemical Warfare Service. A biochemist of that service

found an enzyme in the liver of the rabbit which tears
DFP apart. Now, DFP does not exist in nature and undoubtedly it never has. If it were not for Hitler, no rabbit
might ever have made the unpleasant acquaintance of
DFP. Yet rabbits have the enzymes to dismember this rare
poison.

There is other evidence that these so-called detoxica-
tion processes are performed at random. A poison sometimes becomes *more* toxic after the enzymes finish their
alterations of it. The fate of ethylene glycol, the poisonous
solvent for sulfanilamide (see page 82) is a good example
of this.

Ethylene glycol (the reader may know it as an anti-
freeze), is an alcohol. Many alcohols are poisonous—even
ethyl alcohol, which is the least poisonous of them, can be
quite dangerous. Let us digress a bit from the fate of ethylene glycol and follow the course of ethyl alcohol in the
body.

Ethyl alcohol can reach fatal concentrations in the body
from the rapid intake of about one tenth of an ounce of 200-
proof alcohol per pound of body weight. In other words, if
a man weighing 150 pounds drinks rapidly 15 ounces of
200-proof alcohol, his chances of recovery from his alcoholic stupor are mighty slim. (A variety of factors such as
the state of the subject's health and his previous experience
in the consumption of alcohol make the exact outcome of
the experiment unpredictable.)

Alcohols produce their toxic effects by inhibiting the
rate of respiration within the cells. The assortment of symptoms which mark a man as being intoxicated can be induced
in the most righteous teetotaler, without the use of a drop
of alcohol, simply by reducing the concentration of oxygen

in the atmosphere he breathes. The scarcity of oxygen will limit cellular respiration and the external symptoms of reduced respiration are the same whether it is brought about by an inhibitor or by diminished supplies of oxygen.

The influence of low oxygen concentration has been studied extensively since it is an important factor in the efficiency of pilots—unless, of course, they breathe bottled oxygen. (At 12,000 feet the oxygen in the atmosphere is one third of what it is at sea level.) The pioneer in these studies, the English physiologist Joseph Barcroft, reported that on journeys to high altitudes he has witnessed emotional reactions similar to those experienced after an overdose of alcohol: depression, apathy and drowsiness or excitement and joyfulness, and general loss of self-control. "A person may sing or burst into tears for no apparent reason or be extremely quarrelsome, indolent, and reckless." During the Second World War, some members of the crews of high-flying bombers would take off their oxygen masks on the return trip, for a quick, hangover-less bender.

The fate of all alcohols is the same in the body. They are gradually oxidized. In the case of ethyl alcohol the intermediate stage during the course of the oxidation is the formation of acetic acid, a normal constituent of the body. Acetic acid can be either oxidized further to carbon dioxide and water or utilized as a brick for the assembly of a number of more complex body components. Alcohol, therefore, is really a food. A very limited food, to be sure, since, like sugar, it lacks proteins, vitamins, and minerals.

Now, just as ethyl alcohol is first oxidized to an acid, so ethylene glycol is also oxidized to an acid. But this acid, oxalic, happens to be a merciless poison. The enzymes which accomplish this oxidation doom the animal to quick

death. In this case it would be much better for those enzymes to lie low and do nothing to the ethylene glycol. The
animal might be able to excrete it slowly through the lungs
and kidneys and thus, if the dose is not too large, survive.
Oxidation of even much smaller doses brings certain death.

The struggle against our bacterial enemies seems to be
more purposeful than the disposal of simple toxic substances. However, even this apparently planned campaign
may be really more haphazard than it appears. Bacterial
poisons are either proteins or other complex molecules. We
are unable, with our present knowledge, to view the intimate mechanisms that accomplish the disposal of these
poisons. We only see the ultimate effect and that seems
mighty purposeful.

Those parts of the body which are not normally in contact with matter from the outside world—the blood and
various other tissues—are free from bacteria. Not statically,
the way the inside of a can of evaporated milk is bacteria-
free, but dynamically free. If any bacteria penetrate into the
inner tissues they are attacked by special shock troops for
aggressive defense—the white cells of the blood. They are
smaller in number than the red cells and, as their name
implies, they are without the red hemoglobin. They have,
instead, other specialized equipment—enzymes. The white
cells can project tiny strands of their tissues and encircle
the bacteria. Once trapped the bacteria are helpless, for the
potent enzymes oozing out of the white cells tear them to
shreds.

A boil on the neck is a typical battleground in such a
struggle against marauding enemies. Some microorganisms
lodge near the root of a hair and, finding a warm, cozy nook
and a source of food, begin to multiply. As a first reaction

to this breach in the body's static defense, the area becomes inflamed. The capillaries become dilated, causing the flow of an augmented supply of blood. Fluids from the blood escape into the area and form a clot, converting the whole into a jellylike mass. A barrier of fibrous tissue is formed around this mass, isolating the infection. Meanwhile the white cells have been gathering on the battleground and, crawling through the capillary walls, they pounce upon the infecting organisms. If all goes well, the invaders are killed off in this melee. If not, the infection spreads and the battle is repeated at every new focus of infection. The white cells carry out Mr. Churchill's magnificently expressed strategy: wherever the germs may travel in the blood stream, they are pounced upon and engaged in mortal combat. The white cells get powerful assistance, once the infection becomes very widely spread, from the large phagocytes in the capillaries of the liver, spleen, and bone marrow. (The same cells which dispose of the aging red corpuscles.) If the agencies already committed to the defense are inadequate and the bacteria are winning in those scattered engagements, the body has still another line of defense—but not against every type of invading organism.

It has been part of man's knowledge for a long time that there are diseases from which a survivor rarely suffers again. Apparently that first attack leaves him with a receipt that he has paid his tax of pain to life and he is spared further visits from the same tax collector.

Induction of a mild case of smallpox in order to gain immunity to a severe attack was practiced a score of centuries ago by medicine men in India. They obtained pus from a patient with a mild case of the disease and smeared it into a scratch on a healthy person. The practice of such preventive inoculations against smallpox continued in the East but was introduced into Europe only in 1718 by Lady Mary Wortley Montagu, the wife of the British Ambassador in Constantinople. She had her son inoculated by a Turkish doctor. The wide prevalence of the disease spurred others to follow the example set by the courageous ambassadress and the practice of inoculation against smallpox became widespread. (During the Middle Ages, to be free of pox marks was considered a mark of beauty.) Little wonder that people were ready to subject themselves to the inoculations, even though the outcome was not always predictable. Sometimes the induced disease was a severe case of smallpox.

Chance taught us a safer and equally effective mode of protection against this disease.

Cows, too, are susceptible to smallpox. English farmers knew that cowpox was contagious among humans and considered the rather mild disease as an occupational hazard of dairymaids and others who came in close contact with cows. Who made the first observation that a case of the mild cowpox protects the sufferer from the virulent human smallpox is not certain. Some English farmers have been credited with this astute and profoundly useful correlation. But debates on priority, whether of ancient or of current discoveries, are unprofitable. The discovery is of far greater importance than the ego of the discoverer. Certainly it was Dr. Edward Jenner who established the facts with well-documented evidence and demonstrated how to harvest the benefits of the chance discovery. Since cowpox was known technically as Variolae Vaccinae, the purposeful inoculation with cowpox pus to immunize against smallpox, came to be known as vaccination.

Vaccination became a very widely applied and gratifyingly effective measure of protection against smallpox. But the mechanism which produced the immunity of course remained unknown. In those pre-Pasteur days even the cause of infectious diseases was unknown. Pasteur demonstrated that some diseases and the immunity to them were induced by the same infectious agent. He did this in a celebrated public, scientific demonstration.

Anthrax was decimating the cattle herds of Europe. Pasteur traced the cause of the disease to the tiny rod-shaped bacilli which were teeming in the blood of the diseased farm animals. He also found that these bacilli could be rendered less virulent by keeping them, for a while, at temperatures much higher than normal body temperatures. (Most bacteria have become so accustomed to the warmth of the animals they inhabit that they live best at that temperature even though they themselves are unable to maintain such warmth.) The heat-treated germs would not kill the animals when injected. They merely made the animals ill. However, after the animals recovered from their mild attack of anthrax they were immune to the virulent organisms. They could shake off doses of anthrax bacilli which would surely kill unimmunized animals.

When the conservative physicians scoffed at Pasteur's claims he arranged a public demonstration which had all the trappings of a country fair. Scientists, physicians, dignitaries, and newspapermen all gathered in a field where fifty sheep, a few bottles of bacterial broth, and Pasteur were the center of attention. His assistants injected twenty-five sheep with a heat-weakened bacterial suspension. Twelve days later those same animals received a stronger dose of infection, consisting of bacilli which were exposed to less rigorous heat treatment. The sheep survived this injection, too. Finally all the animals, including the twenty-five "controls" which until then had been untouched, were injected with the same deadly dose of anthrax. Two days later, twenty-two of the controls were dead and the three others were in their final agony. Every one of the sheep protected with the heat-weakened bacilli was grazing contentedly!

Pasteur thus proved not only that infectious diseases are caused by microorganisms but also that those very germs rouse the animal to rally to its own defense and make weapons to repel future invasions. (Of course, today we know that all infectious diseases are not caused by microorganisms. Some, including smallpox, are caused by viruses. But only several years after Pasteur's death were viruses discovered. Still later it was found that microorganisms themselves have infectious viral enemies, the bacteriophages. This finding was made, appropriately enough, by Dr. F. H. d'Herelle, a bacteriologist working in the institute that Pasteur founded.)

The nature of the weapons of immunity remained obscure for a long time after Pasteur's dramatic success with the attenuated anthrax bacilli.

The allergies which bedevil so many of us are also our saviors from germs. Outbreaks of hives, sneezing, the necessity for abstaining from certain foods, are relatively small prices to pay for having the weapons with which to combat infectious disease. For the same process which visits upon us the discomforts of allergies brings to us the blessings of immunity to disease.

Allergy means simply an "altered reaction." If we inject into a guinea pig some egg white from a hen's egg nothing

unusual will happen. But if after some time we inject into
the same guinea pig an identical dose of egg white there
will be a violent reaction. The animal's breathing will become labored, it will thrash around, and will finally go into
shock, from which—depending on the doses injected—it
may or may not recover. The guinea pig has an altered reaction or an allergy to the second injection of egg white.
What brings about this violently different response to the
second injection?

Animals resist the entry of foreign proteins into their
tissues. The proteins we eat do not normally enter our tissues intact. These proteins are broken down by the enzymes
of the alimentary canal into their component amino acids
and only these are permitted to enter our tissues. From the
absorbed amino acids we fashion protein molecules in the
image of our own proteins. (In patients who suffer from
the various allergies there seems to be a minute amount of
seepage of the foreign proteins into their tissues.) Therefore if a foreign protein does enter into the tissues of an
animal it means that a stranger is within the gates. The
stranger may be just a foreign protein molecule or a whole
organism with its foreign proteins. The reaction of the animal is the same to either danger. It begins to fashion shock
troops, or antibodies, in an attempt to dispose of the invaders.

Antibodies dispatch the invaders to their doom in a variety of ways. There are some antibodies which dissolve
the cell walls of bacteria and the tiny monsters just ooze
away. Others merely stimulate, by their very presence, the
white cells to greater efficiency. Still other antibodies combine with the intruding protein or germ, the so-called antigen, to form with it an insoluble particle. Once the foreign

matter is thus clumped together the scavenging white cells and large phagocytes dissolve them at their leisure.

For example, in the blood of the guinea pig which has been injected with the egg white there appears an antibody which when added to a solution of fresh egg white curdles it. The violent symptoms of the guinea pig on the second injection are caused by the excessive curdling between egg white and antibody within the animal.

All antibodies are amazingly specific. The antibody from the blood of this guinea pig will not curdle the egg white from a duck egg or a goose egg as effectively as it does the hen's egg white. It is easy to see why such remarkable specificity of antibodies is essential. An animal would be in a dire predicament if it produced antibodies which curdled any protein at random. The antibody might curdle the animal's own proteins.

That antibodies are so specific is a great asset. We can take the antibodies formed against diphtheria by a sturdy horse and fortify a human child with those very antibodies. Or we can tell whether a brown stain on a cloth is ox blood or human blood. When the dissolved stain is mixed with the serum of a guinea pig which had previously received injections of human blood, the formation of a precipitate identifies the stain as human blood too.

Our practical knowledge of immunochemistry, for that is the name given to the study of antigen-antibody reactions, fills books. But our knowledge of *how* an antibody is formed is nil. We are handicapped by our lack of knowledge of the exact structure of the protein molecule. For within that structure must lie the secret of how one protein molecule, the antigen, can stimulate the shaping of another protein molecule, the antibody, so that united they form an insoluble curd, while each of them separately is perfectly soluble.

Since the study of such properties as solubility has been staked out by physical chemists, some of them were tempted, about a dozen years ago, to study antibodies. Applying their training and knowledge gained in the physical realms, they were ready to polish up problems which have baffled biological scientists for generations. But their contributions, so far, have not been impressive.

Every now and then men well trained and accomplished in physics or chemistry, or just mathematics alone, become impatient with the slow fumbling ways of the biological scientist and decide to come to our rescue. Immunochemistry, cancer, the baffling machinery of the brain, are some of the peaks which intrepid physical scientists have set out to scale recently. (Often, the only reminders that physical scientists have been scrambling at the base of those forbidding peaks is the freshly carved nomenclature they invariably leave behind.) There are more things in the universe that is a living cell than the physical scientists have ever dreamed of. Isotopes, tools of the physical chemists, solved one tiny bit of the mysteries of antibodies. But the tool was wielded by biochemists.

We have seen earlier that the body is in a constant state of flux. Tissues are broken down and rebuilt constantly. The amino acids which compose our proteins today, will be gone tomorrow and replaced by more recent amino-acid arrivals from our food. Are antibodies an exception to this constant building and dismantling? They are known to remain in the body for years after the infection which caused their formation has subsided. Indeed, sometimes they remain for a whole lifetime. Are antibodies permanent islands in the constantly changing sea of the body?

Dr. Michael Heidelberger, the foremost immunochemist, answered this question in collaboration with the late Dr. Rudolph Schoenheimer. They found that antibodies are no different from other proteins of the body. They, too, are being constantly assembled and dismantled. In other words, the machinery which is set into motion at the time of the original infection to produce antibodies, continues its production, sometimes for years, sometimes for a lifetime. We know then that the machinery keeps functioning. But what that machinery is, we have no idea.

Although the writer has emphasized the unsolved problems in immunology, actually the practical achievements in this field are enormous. The multitude of different immunizations, the classification of human blood into various types to insure safe blood transfusions, the diagnosis of diseases (such as syphilis) by testing a few drops of blood, the whole branch of medicine treating the allergies are the bounties harvested from the studies of the antigen-antibody relationship. But the discussion of those achievements can be left to others. To some of us the unknown is far more fascinating than the known.

Bacteriological warfare is a new term, much bandied about lately. It refers, of course, to the use of bacteria or their poisons as weapons in human warfare. But bacteria, too, have been carrying on warfare on a vast scale, millions of years before man and his puny wars became part of the earth's landscape. Man is a fumbling novice in the techniques of mass slaughter. What is it, a hundred thousand humans that we can kill with one atomic blow? In less time than it takes to say "hydrogen bomb," millions of bacteria are the casualties in the warfare that goes on in a handful of garden soil. And the weapons used are at least as ingenious as man's.

In that handful of soil are more microorganisms than there are humans on the face of the earth. Life is hard in that handful of a world. Food is scarce; the struggle for survival is fierce. Some of the combatant microorganisms are especially well equipped in their battle against their competitors. They pour out a poisonous solution all around them. In the area staked out by the spreading molecules of the poison, no other organisms can enter and live. The wielder of the poisonous weapon can grow and multiply in his befouled homestead.

It is almost superfluous to say that it was Pasteur who first observed this Lilliputian chemical warfare. A batch of the anthrax bacilli which were described a few pages back stopped growing. He pinned the responsibility for the mass murder on some stray microorganisms which drifted into his cultures from the air.

In the garden of the mind of that genius, this chance observation became the seed of a dream. He visualized the slaughter of the pathogens which cause our diseases by the introduction of their natural enemies or their products into the patient. He wrote in 1877 that such a scheme "justifies the highest hopes for therapeutics."

It has taken sixty years for that dream to come true. There were little episodes which kept the dream alive during those years. In 1885 Cantani tried to cure a tuberculous woman by the inhalation of a nonpathogenic organism. He was apparently fairly successful. (It may have been a case of a normal recovery from tuberculosis.) At any rate the work was not followed up.

However, many bacteriologists continued to notice the lethal antagonism of some microorganisms to each other. The disappearance of pathogens from water which trickles through the soil became well known.

The destruction of one microbe by another came to be known as antibiosis. At the start of this century the first antibiotic—a substance extracted from one microorganism to kill another—was prepared. However, it was not successful.

The credit for the first isolation of an effective antibiotic goes to Dr. René J. Dubos, a bacteriologist, of the Rockefeller Institute and to his biochemist associate, Dr. R. D. Hotchkiss. Unfortunately their antibiotic, while very effective against bacteria, is also quite poisonous if injected into the patient. It is not used widely, except for surface applications, nor has it received the wide public recognition it deserves.

Dubos' work is the model in the search for all antibiotics. He took a pinch of soil and sprinkled it onto a glass dish coated with nutriment for bacteria. He separated the various strains of bacteria which grew and replanted them into little bacterial gardens of their own with lots of food. Having feasted them, Dubos put the bacteria to work. He placed a growing culture of pathogenic organisms—staphylococci, which cause boils—into each colony of soil bacteria. One strain of the soil bacteria refused to share its food with the staphylococci. These bacteria exude something which kills competing organisms.

The isolation of an antibiotic, later named gramicidin, followed the usual pattern of such searches. The poison-

bearing bacteria were grown in large batches and the antibiotic was concentrated from their juice, using the increasing toxicity of the preparations to the staphylococci, as the
guide in the various steps of the search. The task was completed in 1939. However, the antibiotic turned out to be
quite poisonous when injected into animals. Nevertheless,
it is a valuable aid in the treatment of exposed infections.
But there were other, better antibiotics to come.

Streptomycin was dug out of bacteria by Dr. Dubos'
former mentor, Dr. Selman Waksman of Rutgers. This is a
rare case of the master following in the footsteps of his
disciple.

Waksman has devoted his life to the study of soil microorganisms. They are of enormous importance agricultur-
ally, economically, and even aesthetically. About a ton of
leaves falls on each acre of forest every year. If it were not
for the soil organisms which decompose all such debris,
our earth would become a cluttered, uninhabitable grave-
yard in a very short time. No dead plants or animals would
decompose. The substance of their bodies could never be
returned to usefulness. Our accumulation of ancestors for
thousands of years back would be with us perfectly pre-
served.

Waksman had been interested in soil organisms from
the point of view of their value in agriculture. But after
Dubos had extracted an antibiotic from such organisms,
Waksman, too, channeled his efforts in a similar direction.
Streptomycin is but one of the many antibiotics he and his
associates extracted.

And now a few words about the greatest antibiotic of all
those that have been isolated thus far—penicillin. In 1928
a mold spore drifted from the air onto a bacterial culture

plate of Dr. Alexander Fleming, a bacteriologist at St. Mary's Hospital in London. The spore reproduced on the spread of food in the glass dish. As it grew, it exuded something, for there was a halo around the mold, an area free of the staphylococci which were the original inhabitants of that plate. Now, such accidents must have happened to scores of bacteriologists. But they would simply throw a ruined plate into an antiseptic bin.

Fleming had been carrying on bacteriological warfare—the proper kind, *against* bacteria—for a quarter of a century. He became interested in this mold which could parachute into a colony of pathogens and could slaughter them with ease. He transferred the mold to culture plates and grew it in large batches. The mold-free juices he prepared were still lethal to bacteria. Since the mold was a *Penicillium* Fleming named the antibacterial substance in the juice penicillin, after the parent. Fleming recognized immediately the potential value of penicillin. He wrote in 1929, that "It may be an efficient antiseptic for application to, or injection into areas infested with penicillin-sensitive microbes." But, as he wrote more recently, ". . . I failed to concentrate this substance from lack of sufficient chemical assistance. . . ."

Not until ten years later was penicillin concentrated and used in human therapy. Dr. H. W. Florey, a pathologist at Oxford, undertook in 1938 a systematic search for antibiotic substances. His biochemist associate was Dr. Ernst Chain, a refugee from Hitler. The team of Florey and Chain was spectacularly successful in winning penicillin. Their methods were communicated to American laboratories and pharmaceutical houses under the auspices of the Office of Scientific Research and Development. "The Americans," wrote Fleming, "improved methods of production so that on D day there was enough penicillin for every wounded man who needed it. . . ."

There is beautiful, fairy-tale justice in Chain's career. Exiled from his home, he became a key man in the fashioning of a drug which protected millions of young champions as they sallied forth to liberate his homeland.

Why was there a lag of ten years between the discovery and the perfection of one of the greatest drugs? Was it Fleming's fault? Certainly not. He was trying to enlist aid and interest; he sent filaments of the mold to any one who wished to grow it and study it. Was it the fault of scientists in general? No. As we shall see later, they can hardly be blamed. Who then is to be blamed that during those ten years men perished by the tens of thousands from infections which, with the aid of penicillin, they could have conquered? But let us leave this to the last chapter where the lag of ten years will be the leitmotif of a discussion on the neglect of research.

How does this most potent of cell defenses—which we borrow from molds for our own defense—kill bacteria? Oddly enough, we are still not certain. There is a tentative explanation, but it leaves much unexplained. Penicillin is known to inhibit some enzyme systems in susceptible bacteria. Those particular enzymes may be absent from the organisms which do not succumb to penicillin. The absence of such enzymes from animals may also be the reason for the nontoxicity of penicillin to humans.

We know the chemical structure of penicillin and have even made it in the laboratory. How simple that sounds!

That brief sentence summarizes years of work during the Second World War by scores of the best organic chemists of England and America. This Allied team was led by two of the ablest generals of organic chemistry of the two countries: Sir Robert Robinson, who came to this campaign after brilliant victories in the field of the structure of natural pigments, and H. T. Clarke, who decoded the structure and achieved the synthesis of the sulfur-containing part of vitamin B_1.

While the synthesis of penicillin is a brilliant achievement of this international team, so far it has no practical significance. It is cheaper by far to let the molds do their chemistry and make it for us.

We do not know the specific function of penicillin within the mold itself. We know altogether very little about the biochemistry of microorganisms. With excusable narcissism we have concentrated so far on the study of human biochemistry or of animal biochemistry related to it. Penicillin may be a normal metabolic product of the mold which happens to be poisonous to other microorganisms; or, it may have no role within the cell other than self-defense. A chance mutation, resulting in the ability to make penicillin, may have enabled that particular mold to survive, for, nature achieves her purpose in many apparently purposeless ways.

The parallel history of penicillin and of Dubos' antibiotic illustrates the role of chance in the rewards of research, too. Two groups of scientists, Fleming, Florey, and Chain on one hand, and Dubos and his team on the other, set out on a hunt for an antibiotic like so many prospectors hunting for gold. The prospectors have the same training, the same skills and tools. The team which succeeded first, later found fool's gold mixed with the gold: the antibiotic was toxic. The other team, by chance, found pure gold, penicillin, and received accolades and Nobel Prizes. But, well has it been said that "The object of research is the advancement, not of the investigator, but of knowledge."

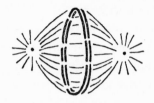

The blueprint of our cells

"Glory to the great friend and protagonist of science, our leader and teacher, Comrade Stalin!" With this fulsome dedication ends the book *The Science of Biology Today* written by Trofim Lysenko, the winner in a gruesome struggle among Soviet biologists. The outcome of the struggle was not too difficult to foretell. Comrade Stalin was in Lysenko's corner or, to be more accurate, Lysenko was in Stalin's corner.

What the defeated biologists think of the "great teacher," is difficult to report with accuracy; *their* books must await the time when the ultimate in Five Year Plans brings better publishing facilities to Siberia.

The apotheosis of Mr. Stalin as the number one biologist of all the Russias is a logical development which merely repeats a compulsive pattern in the relation of science to all dictatorships. Under a totalitarian atmosphere science wilts and the scientists themselves shrivel, or worse. Since the *sine qua non* of the scientist is a free, inquiring mind, which, in turn, is the strongest barricade against enslavement, the scientist becomes the public-enemy-number-one of all totalitarian regimes. Galileo suffered the ignominy of recantation, Roger Bacon languished in jail, and Giordano Bruno was burned at the stake because they challenged the authoritarian precepts of their time. The fate of the scientists under Hitler and Mussolini is still all too painfully vivid in our memories. And in Russia, scientists must step behind that great teacher, Comrade Stalin.

The crux of the controversy in which Lysenko emerged victorious is simply this: Can the characteristics acquired by a living thing in its lifetime be transmitted to its offspring?

It is at first surprising that an academic question which has been debated for about sixty-five years should become a political issue. Three factors raised the moot problem in biology to an affair of state: According to Marxist theory a change in man's political environment could purge him of his base acquisitive instincts.[1] It is doubtful that Stalin would be aware of such academic questions in biology if ambitious biologists had not brought it to his attention. And finally, the totalitarian state cannot tolerate the slightest deviation from the true path. It might become contagious.

The mechanism of heredity which has fascinated biologists for generations, has recently been partially revealed. Needless to say, it is a chemical mechanism. But before we unveil this latest masterpiece shaped by the tools of biochemistry let us review the groundwork of generations of biologists which forms the pedestal on which this achievement stands.

That the gross, species-characteristics of the parent are

---

[1] The ennobling Soviet atmosphere has as yet made no dent in the self-promoting instincts of these biologists.

handed down to the offspring through the seed has been obvious to man for scores of centuries. The farmers of antiquity, planting their grains along the Nile, the Euphrates, and the Yangtze, confidently expected to harvest the same kinds of grains.

That the variations among individuals of the same species can be handed down to the offspring too, was also part of man's knowledge for a long time. The Arabian racehorse breeders understood the importance of heredity in improving the speed of their horses so well that they kept elaborate pedigree records for centuries. Indeed no race horse could be labeled a purebred Arab unless its ancestry could be traced to one of five famous progenitors of the Arab breed.

While it was recognized that certain traits are passed down from parent to offspring, the patterns—indeed the laws—which heredity follows remained unknown until the middle of the last century. Even then their discovery remained obscure and the announcement of the discovery itself had to be rediscovered thirty-five years later.

Gregor Johann Mendel chose for himself a life of peace and obscurity as an Augustinian monk. But for his hobby, he would have remained one of the anonymous brothers of the monastery at Brunn, Austria. However, he pursued his hobby in the tiny monastery garden with such brilliance that he was lifted from his self-imposed anonymity and he is counted among those men of rare genius who are the first to discern a law of nature. Mendel was the founder of the science of genetics. He shaped the tools for the study of the ways of heredity.

He studied the effects of the crossbreeding of two plants endowed with contrasting traits. He started by planting different peas, tall and short, in the monastery garden. If these two varieties of peas were allowed to self-fertilize, with no possibility of cross-fertilization, the plants yielded seeds which bred true to type: they grew into plants, tall or short, like their parents. But if a tall and a short pea were cross-fertilized or "hybridized," the seeds from such a union gave rise only to tall peas. Mendel did not stop there. He patiently crossed these tall peas of mixed ancestry and collected *their* seeds. Out of these seeds grew both tall and short peas. He carefully recorded his seeds and crops and found from over a thousand different plantings that the tall and short "grandchildren" always appeared in a definite ratio: 75 percent tall and 25 percent short. He repeated these studies with red- and white-flowered peas. When he cross-fertilized these flowers he found that all the seeds gave rise to red flowers. But these second generation red flowers, when cross-fertilized, produced seeds from which grew both red and white peas. He found 6,022 red and 2,001 white flowers, again a ratio of three to one (75.1 to 24.9 percent, to be more exact).

Mendel concluded that there are factors in peas which determine their color and height. The factor for whiteness or shortness remains dormant in the first generation after the crossbreeding of opposing traits, but asserts itself in the second. Moreover, the dormant factor reappears in the second generation in a ratio of one to three of the dominant factor.

This profound discovery lay hidden for thirty-five years in the pages of the local scientific journal in which Mendel published it. The eyes of the whole scientific world were finally focused on the great discovery when in 1900 three different biologists simultaneously stumbled on the life work of the patient monk. Building on this foundation, other biologists quickly erected the science of the study of heredity, genetics.

The next great experimental approach was made by Dr. Thomas Hunt Morgan, who undertook to unravel the hereditary maze of an insect, the fruit fly. He found that the fruit fly has a great many traits inherited in a Mendelian way. Some of these traits tend to appear together and four such groups of linked traits could be identified. It was known that in the cells of the fruit fly there are four pairs of rodlike structures, chromosomes (colored bodies), whose function, however, was a mystery. Morgan surmised that these chromosomes might be the seat of the machinery which hands down Mendelian traits from generation to generation.

Since there are more traits than chromosomes and, since several traits are linked together, the conclusion was drawn that one chromosome houses the controlling mechanisms for more than one trait. These tiny, invisible fragments of the chromosome, each of which controls a single trait, were called genes. It is believed that a separate gene exists for every single trait.

The larger aspects of the mechanism of heredity now became clear. Every individual has two sets of chromosomes, one set contributed by each parent. We humans have twenty-four pairs; twenty-four from each parent. (The reproductive cells contain only one half as many chromosomes as the other cells, otherwise our chromosomes would be doubled every generation.)

To illustrate the mechanism of heredity let us return to the sweet pea. When a white and a red sweet pea are cross-fertilized, each sex cell contributes a gene for the control of color; one a red, the other a white. For some reason the red gene dominates over the white and such a hybrid seed will always give rise to red flowers even though it contains both the red gene and the white gene. In forming the sex cells of this hybrid plant only one of these genes, the red *or* the white, goes into each sex cell. Thus these sex cells contain the red or white gene in the same ratio, one to one. The probabilities of fertilization are as follows: A red male can fertilize a red female; the offspring from such a union will be red because both genes are red. A red male cell can fertilize a white female; the offspring from such a union will also be red because the red gene is dominant over the white. A white male can fertilize a red female, producing again a red offspring. Finally a white male can fertilize a white female producing the only white offspring of the four. This is the reason for the three to one ratio of red to white peas that Mendel recorded.

As far back as 1883 a German biologist, who was unaware of Mendel's work, differentiated between traits acquired by an individual during its lifetime and those which are handed down to it from the cells of its parents. Weismann thought that the acquired traits cannot be passed on to the offspring. To test his assumption he subjected experimental animals to drastic environmental changes. For example, he cut off the tails of mice for nineteen generations. He could observe no change whatever in the tails of the new-born mice from the mutilated ancestors.

It was known, however, that from time to time some novel traits which *are* passed down to the offspring appear in individuals. These freaks or sports appeared here and there by rare chance. To some of the inhabitants of a certain Swiss valley six-fingered babies are born; among the orange trees of Brazil a seedless, navel orange makes a sudden unexplained appearance; a silver fox gives birth to semi-albino offspring and launches the platinum fox industry. The biologists could but wait for these sports or mutations to occur; they could do nothing to induce them. They tried valiantly despite Weismann's failure. All kinds of environmental changes were devised to induce such inheritable mutations but to date there has been no unequivocal achievement of it.

Finally, biologists resorted to stronger measures. Dr. H. J. Muller, one of T. H. Morgan's students at Columbia, exposed the fruit fly to X-ray radiation. All kinds of freaks appeared among the progeny of these flies; extra wings, a lack of wings; extra legs, no legs. Biologists were delighted. The gene was approached at last. Apparently the intense energy of the X ray when it hits the gene target dislocates its structure, producing internal havoc and monstrous offspring.

The X-ray-induced mutations threw a light on the possible causes of natural mutations. Plants and animals are exposed to all kinds of potent natural radiations. The radioactive minerals radium, uranium, and thorium give off high-energy radiations as a flower gives off its scent. (Since the radioactive elements keep disintegrating at a constant rate we know that the quantities of such elements on earth today are but a minute fraction of the amounts that must have been here eons ago.) Cosmic rays are constantly bombarding us. Our earth is bathed in ultraviolet rays. A natural mutation may very well be induced when a gene in a sex cell of a plant or animal happens to get into the path of a packet of energy from any of these sources. However, this is still in the realm of speculation; the spontaneous mutations may be induced by stresses, other than radiation energy, within the chromosome. One current theory is that they are produced chiefly by random collisions between molecules.

Muller's man-induced mutations eventually enabled us to study how the tiny submicroscopic gene controls the shape, the color, the very essence of a living thing. A fortunate partnership between a geneticist and a biochemist was formed to explore this problem. Dr. G. W. Beadle and Dr. E. L. Tatum of California studied the mutations of a very simple organism, the bread mold *Neurospora crassa*.

This little mold makes very few demands on the world. It thrives if it gets sugar, water, some salts, and just a dash of one of the B vitamins, biotin. Out of these few starting materials the mold fashions an astonishing variety of substances for itself. It makes its own amino acids, its own vitamins, indeed anything else it needs for life. It owes its versatility to the large variety of enzymes it has for such tasks of synthesis. For every chemical process in living things is enzyme operated.

But if this self-sufficient mold is exposed to X-ray irradiation some of the next generation can no longer live on the simple diet of the parents. These offspring will die on a diet of sugar, salts, and biotin. Some will need one or more of the vitamins, others will need some amino acids. Only when these foods are provided will these crippled offspring live. And the dependence on these extra food rations is now handed down from generation to generation, proving that a mutation has taken place—a mutation which made the mold lose the know-how of making amino acids and vitamins. Since it was enzymes which made these foods in the natural *Neurospora*, we can conclude that these particular enzyme potencies must have disappeared under the impact of X rays. But these traits, which are transmissible from generation to generation, reside in the genes; therefore the function of genes becomes apparent: they control the fashioning of enzymes.

Before these brilliant discoveries were made we had hints that the genes control our enzymes. There is a rare type of idiocy among humans called oligophrenia phenylpyruvica. The unfortunates who are born with this affliction all have tiny skulls (oligophrenic); their intelligence quotient is anywhere from 20 to 50; and they are always very blond. These patients cannot metabolize one of the essential amino acids, phenylalanine, completely. Since the body's pigments are made out of this very amino acid, the reason for the extreme blondness of these unfortunates becomes apparent. One of the intermediate products of the metabolism of this amino acid can be found in the blood and in the urine of these patients. Just as diabetics void sugar because they cannot handle it, these subjects excrete a product of amino acid metabolism with which they cannot cope.

The disease is definitely hereditary. Some ancestor of these patients lost the ability through a ghastly mutation, to make a particular enzyme in the chain of enzymes which metabolize this amino acid. Since both synthetic and degradative processes are performed through a series of enzyme-motivated steps, if but one such a step is knocked out in the cell's assembly line, the synthesis—or the degradation—of a particular substance is stalled. Such enzyme-endowed abilities were gained and lost apparently at random, through the long evolutionary history of a present-day organism. There is no correlation between the position of an organism on the evolutionary scale and its synthetic ability. Some microorganisms, like the bread mold, have prodigious capabilities in fabricating amino acids and vitamins; others—for example, the lactic acid organisms—must obtain dozens of preformed essentials from their environment or they starve to death. Even among so homogenous a class as the mammals there are wide discrepancies in synthesizing ability. The white rat cannot synthesize the amino acid histidine, but another member of that class, man, makes this amino acid with ease. There is recent evidence to indicate that the ability to fabricate a given compound might have been gained and lost several times during the evolutionary history of an organism. It is most probable that every synthetic ability was acquired by mutations. The chances are that the first ancestral living forms had *no* synthesizing abilities whatever. It is difficult to visualize how such primordial forms could have arisen *de novo*, equipped with the complete battery of enzymes needed to synthesize the compounds they required for life. The likelier assumption is that they could but forage, as it were, on the multitude of carbon compounds which must have been scattered in generous abundance on the cooling crust of our once hot planet. The ability to synthesize must have accumulated as the enzymes—or, rather, the genes for the production of those enzymes—were slowly acquired by fortuitous mutations. The source of the staggering biochemical complexity of contemporary organisms now becomes clear. A microbe or a man, is a biochemical potpourri, a summation of those haphazard mutations.

There have been complaints from lay critics that the biological scientist is barren of imagination. He is accused of rarely, if ever, making the sweeping generalizations of, for example, the atomic physicist. The fault, if it is a fault, is not the biologist's, it is nature's. Either there is no over-all pattern to life, or, if there is such a pattern, it is much too complex to be revealed to our current, undeveloped minds. How can we predict whether a particular enzyme is, or is not, present in a living thing which is here today as the culmination of thousands of haphazard mutations? In man's blood and urine we find a substance, uric acid, as a waste product of protein metabolism. In dogs, uric acid is almost completely converted by an enzyme in their livers to something else, allantoin, which they then excrete. How-ever, one breed of dogs, the Dalmatian coach dog, lacks this particular enzyme and these dogs excrete not allantoin but uric acid, as man does. We challenge the greatest minds to tell us how we could have predicted a priori that the Dalmatian coach dog would follow the human pattern in respect to this waste product.

The knowledge of the biological scientist comes from cumulative experimental probings. He must patiently un-fold the complexities of the cell, he can rarely predict them. This explains the dearth of infant prodigies among out-standing biological scientists. The mastery of the tech-niques of his craft and the accumulation of knowledge with those techniques takes so long that the biologist is usually well into middle age before he achieves eminence in his field. Not so the mathematicians and theoretical physicists, many of whom are hardly out of their teens when they make their niche-carving contributions. But their task is simple. They have staked out for themselves only the inorganic universe, the atoms, the planets, the stars. The biological scientist surveys the microscopic cell into which are packed all the forces and components of the inorganic universe *plus* the force and pattern of life.

How does a protein molecule, the gene, produce another protein molecule, an enzyme? Or, an equally vexing question, how does the gene duplicate its very self? We humans are launched into life from a single microscopic fertilized egg. This egg contains in its nucleus 48 chromosomes which in turn contain an uncounted number of genes. Eventually every cell in our body contains 48 identical chromosomes, presumably with identical genes. How did the genes duplicate themselves billions of times to populate billions of our cells?

In order to be able to speculate about these questions intelligently an entirely different field, that of the viruses, must be explored first. Viruses are responsible for some of the most devastating diseases of plants, beasts, and man. When we are visited by the appropriate virus we may come down with smallpox, yellow fever, dengue fever, poliomyelitis, measles, mumps, influenza, virus pneumonia, and the common cold. Plants and microorganisms, too, are plagued with a variety of virus infections. The most important of these, for our purposes, is the tobacco mosaic disease. On tobacco leaves a brownish mottling sometimes appears which tends to assume a mosaiclike pattern. The disease can be spread by spraying a healthy plant with the juice from a macerated diseased leaf. About fifty years ago a hunt was started for the infectious agent. Such a hunt is relatively simple if the culprit is a microorganism. The infectious liquid is passed through a very fine filter made of unglazed porcelain. The organisms are held back on the filter and the fluid which passes through is no longer infectious. But the juice from the diseased tobacco was just as infectious after it passed through a filter as it was before. The agent which causes the disease is too small to be held back, it passes through even the tiny, microscopic pores of the porcelain. The name filterable virus was given to the elusive agent.

During the next fifty years a number of other viruses were detected, including the bacteriophages which infect microorganisms. But no viruses were isolated. They were never seen; they were known only by the diseases they produced in plants, animals, and microorganisms.

Finally, in 1935 a virus was tracked down by Dr. W. M. Stanley of the Rockefeller Institute. Nine years earlier Dr. J. B. Sumner had isolated a crystalline enzyme, urease, which turned out to be a protein. Stanley collected the juice from large numbers of crushed, diseased tobacco leaves and started to concentrate the virus not mechanically on a filter, but chemically. He took those steps which were known to concentrate proteins. He could trace the path of the virus during his manipulations by spraying each fraction in his separations on fresh, healthy plants. Eventually he purified the virus to such an extent that it crystallized out just as Sumner's enzyme did. The virus turned out to be a crystalline protein! It had no cellular structure at all. It was a dead protein, no different in appearance at least from crystalline egg white in a bottle on a biochemist's shelf.

When this "dead" chemical was taken out of a bottle and a solution of it was painted on a tobacco leaf it became very much alive. The leaf became diseased; its proteins were converted into virus proteins. The virus is able somehow to take the tissues of the leaf and refashion them in its own image.

There is a very crude analogy to this remarkable phenomenon in the nonliving world. The chemist often encounters solutions from which he is unable to induce the dissolved material to precipitate in crystalline form. But if he drops a crystal of the same substance into the recalcitrant solution the dissolved material begins to come out of solution forming hundreds of new crystals all clustered around the original "seed." Of course this analogy oversimplifies the problem of the formation of virus crystals but it does present a pattern, however crude. The arrangement of the molecules in the seed crystal induces the molecules in the solution to rearrange themselves in the image of the seed. For want of a better analogy, we might visualize the reproduction of a virus protein molecule from its host's proteins along the same lines.

Viruses are unable to reproduce except in intimate contact with living cells. We have not yet been able to induce growth of a virus in any cell-free medium. Viruses are complete parasites. They can but siphon the energy and the substance of their living prey.

Whether synthesis of normal protein or of virus protein takes place in a susceptible, infected cell teeters on a razor's edge. This has been vividly demonstrated by an important recent discovery. There have been conjectures, based on scanty evidence, that there are microorganisms which always have associated with them infectious virus particles. They were supposed to carry, as it were, the seeds of their own destruction. It was thought that they needed no infection from the outside, just the proper conditions for the reproduction of their own indigenous virus particles. Since such organisms would perish sometimes for no apparent reason, they were called lysogenic—self-dissolving. Whether there is a true association between the microorganism and the virus particle, or whether the infectious agent just happens always to be present in cultures of lysogenic organisms could not be decided.

About three years ago, Dr. André Lwoff of the Pasteur Institute in Paris decided to resolve this moot question concerning lysogenic microorganisms. It is technically possible to seize under a microscope a single bacterium and place it in a fresh nutrient medium for culturing. Thus, a pure, homogeneous colony of any microorganism may be started from a single cell. Lwoff selected at random a single cell of a lysogenic bacterium. He placed it in a fresh medium and waited for growth and subsequent cell division. He seized but one of the two offspring and replanted it. With virtuoso skill he repeated this nineteen times. In other words, he had an organism of whose ancestry he was certain for nineteen consecutive generations. Furthermore, it can be calculated that if his inoculum was always one tenth the volume of each fresh culture broth, the original culture broth which might have contained the associated virus had become diluted ten billion, billion fold and thus the original virus must have been lost. Nor could the original virus have reproduced. Viruses can reproduce only within a living cell and they always murder their host as they reproduce. But in this case every ancestral cell for nineteen generations had been accounted for. Yet the bacteria from a fresh colony raised from such a pedigreed cell were still lysogenic: if deprived of oxygen or if permitted to stay in a culture too long they would dissolve and liberate virus particles. Furthermore, Lwoff found that he could induce virus production at will in lysogenic organisms. He discovered that if a fresh culture of such organisms is exposed to ultraviolet irradiation, within a brief time these organisms will be killed by the scores of virus particles which develop in every single cell. In other words, these organisms either have the virus always associated with them or—a more exciting possibility—they may give rise to the virus as the result of some blunder in their own metabolism. In either case, so delicately balanced are the forces for normal growth and for the production of virus that the minute energy of incident ultraviolet irradiation can tip the scale toward the production of the lethal virus particles.

Are viruses alive? The answer depends on the definition of life. Rather than be engulfed in the quicksands of semantics, the writer will evade a direct answer; the reader is as qualified to ponder the question as he is.

The changes in the living world are never abrupt. Between the fish and the land animals stands the lungfish. Between the true mammals and the nonmammalian forms is the Australian duckbilled platypus whose milk pours not from mammary glands but from sweat glands and drips along its hair. The apes separate us from the rest of the animals in the spectrum of life. We might consider the viruses as bridging the gap between the nonliving crystals and the true living forms.

We do not believe, however, that the present-day viruses are remnants of ancestral forms from which other, more complex, living things had arisen. On the contrary, viruses are probably degenerated remnants of previously more complex forms. Viruses are often highly specific parasites: they can "live" only in contact with the cells of specific hosts. If the viruses are remnants of primordial, ancestral forms how could they have survived during the eons before their current hosts, of relatively recent evolutionary vintage, had made their appearance?

What can we learn from the reproduction of viruses

about the multiplication of the genes of a single fertilized egg to populate every cell of the adult with genes? Both genes and viruses proved to be made of the same kinds of proteins, the so-called nucleoproteins. It is attractive to speculate that the genes, too, reshape other proteins in their own image as do their structural cousins the viruses.

There is another demonstration from a different source that the huge molecules of living things can be assembled only if there is a pre-existing molecule to act as a dress-maker's pattern.

Starch is a huge molecule made in plants, such as the potato, by clipping together hundreds of smaller glucose molecules. Dr. Carl Cori and Dr. Gerty Cori, a husband and wife biochemical team, the third couple in history to be honored by a joint Nobel Prize, repeated the starch-making feat of the potato, but they made starch in a test tube. They purified an enzyme preparation which, they felt confident, should be able to join together the hundreds of glucose molecules into starch. All their attempts were fruitless. Finally they added just a trace of starch to their solution of enzyme and glucose and at once the glucose molecules began to be assembled into copious amounts of starch!

It is very probable that the fertilized egg, even the microscopic human egg, contains a few of every kind of protein and other molecules that the adult has. These molecules must be there as patterns for the shaping of the tremendous variety of substances which eventually make up the billions of cells of the adult. Since the molecules are infinitesimally small compared even to the microscopic egg, we can calculate that billions of such molecules can fit even into that tiny speck of protoplasm.

Work on the tobacco mosaic virus revealed the kind of change that probably takes place in a gene when mutation occurs. Viruses undergo mutation, too. While it is impossible to gather enough genes of a plant or animal before and after a mutation to study it chemically, it is relatively easy to do this with the tobacco mosaic virus. There are various strains or mutants of this virus. Each can be separately cultured on different plants and then can be harvested separately. One structural difference between viruses of different strains is already known: the proteins contain different amounts of the constituent amino acids. Indeed, some amino acids present in one strain may be entirely absent in another.

If we assume that mutations in animals follow a similar pattern—a justifiable assumption since nature so often repeats itself—the difference between a musical genius and an ordinary mortal may very well be due to but a slight difference in the amino-acid content of a few of their genes.

And now, at the end of our story of the genes let us return to the Lysenko controversy. Ordinarily such a controversy would be but a tempest in a teapot whose bubbling would not be heard outside of a small circle of scientists.

Controversies in science are not unusual. Two groups of workers can very easily produce clashing fragments of evidence and interpretation as they grope in the intricate maze of a living cell. Nor are such controversies necessarily unhealthy. A discovery of a hidden secret of nature is a joyous event, but some scientists get an extra fillip if such a discovery at the same time demolishes those retarded colleagues who have the temerity to hold opposing views. Sometimes the pursuit of a controversy to a triumphant conclusion is a zestful motive even for the scientist who is completely consecrated to the pursuit of truth. Such controversies are often spirited affairs. Sides are taken by the onlookers; particularly deft thrusts by tongue or test tube are cheered; the contestants are goaded. Only professional dignity is the deterrent to the placing of bets on the outcome.

But when a supreme policy of a totalitarian state is entangled in the controversy it becomes a grim affair. The state cannot afford to lose.

How low the Russian authorities can stoop to bolster their side is indicated by their conversion of an erratic Austrian scientist, a Dr. Kammerer, into a hero of Soviet biology. Needless to say he believed in the inheritance of acquired characteristics. Not only did he believe in it, he demonstrated it. He claimed that he induced changes in the pigmentation of salamanders by changing the color of their environment. (Some salamanders change their color, chameleon-like, enabling them to blend into their background.) Kammerer claimed to have produced permanent, hereditary color changes in his salamanders by keeping them against a background of the same color for several generations. The ingenious geneticist enjoyed quite a reputation until someone multiplied the minimum time needed to raise a generation of salamanders by the number of generations which Kammerer claimed to have studied. Unfortunately for his reputation, it was found that Kammerer's own statement of the length of his studies fell far short of the time required to raise all those generations. More damaging still was the detection, by an American biologist, of

India ink injected under the skin of toads which had been induced by their "environment" to become black.

Kammerer fled, after his exposure, to the Soviet Union, where he committed suicide. According to the great geneticist Dr. Richard B. Goldschmidt there was a movie widely distributed in Russia in which the hero's exploits were patterned after Kammerer's career. The movie hero was framed by bourgeois villains who injected dyes into his experimental animals.

The bizarre aspect of the whole controversy is the tremendous fuss made about an academic argument. No free biologist would be the least bit ruffled if someone should come forth with evidence of a new Mendelian trait achieved by changes outside of the gene. Within the briefest possible time the claims would be tested in a dozen different laboratories and if they were confirmed, we would simply change our pattern of thought about the whole problem; we have changed our concepts before, when facts warranted it. It is not so simple in a totalitarian state where the rulers must save their face, and the scientists their heads. The only permanent effect of the squelching of scientists is the deterioration of science itself. Confirmation of this comes from an unexpected source. That authority on squelching, Herr Goebbels, wrote in his diary:

"Our technical development both in the realm of submarines and of air war is far inferior to that of the English and the Americans. We are now getting the reward for our poor leadership on the scientific front, which did not show the necessary initiative to stimulate the willingness of scientists to co-operate. You just can't let an absolute nitwit head German science for years and not expect to be punished for such folly."

Cells that think

THE IMPACT OF A GENIUS is sometimes detrimental to a field of research. Freud was such an unwitting brake on the development of brain physiology and chemistry. With brilliant insight he penetrated slightly the mists shrouding the abyss of the human mind; his influence on the therapy of mental ills, as well as on our literature and art —not to mention our dinner conversation—is profound. Yet, despite the bold strides with which he led his disciples in the verbal exploration of the human mind, he had an almost paralyzing influence in another area: he made us forget that the brain is an organ.

It is an organ crammed with chemical machinery. It burns foods; it consumes energy; it builds up and breaks down its own tissues. On top of all this, or rather, with the aid of all this, it performs its supreme function: it thinks. The brain is thus an organ of staggering complexity, but it is an organ, no less than the kidney and the liver are organs. Our complaint is that under Freud's hypnotic influence much effort has been focused on the psychoanalysis of the brain and little on its chemical and physiological analysis.

The implication is not that Freud's influence has been bad. There are any number of puny people who try to eke out a bit of fame by associating their names with the great, if only as their detractors. Attack gets more attention than adulation. The writer has no desire to join the ranks of the fifth raters who would achieve fourth-rate stature by attacking a first-rate mind and its works. Indeed, he believes that with no amount of chemical or physiological research using our current, crude techniques could we have advanced as rapidly in the brief space of a few decades as we did following Freud. The two different approaches should have been encouraged to proceed side by side; psychoanalysis for its rapidly reached immediate gains and physiological research for its slow, ponderous, but more profound exploration of the mechanism of the mind.

Of course, in all fairness, it should be pointed out that there were formidable obstacles to the physical exploration of the brain long before Freud. Until the seventeenth century man's soul and mind were considered by many to be akin to a gas. But then, as we began to unveil the mysteries of gases, refuge was taken behind more impenetrable veils of mysticism; Descartes separated mind and matter. He staked us out into two tightly fenced areas: "l'âme raisonnable" and the "machine de terre." He conceived a unique metabolism for the brain. The blood was supposed to induce in the brain "a very subtle air or wind, called animal spirits." The dualism of mind and matter dominated our thinking for centuries. Indeed, it still dominates it today.

That the brain might have an independent metabolism, instead of being a Cartesian windmill, began to be recog-

nized from the work of a remarkable physician who turned
to chemistry, John Lewis William Thudichum. Although
he was born and educated in Germany, he spent most of
his life in England. A physician by training and practice, his
major work was in chemistry; author of the classic *Chemical Constitution of the Brain,* he also wrote *The Spirit of
Cookery,* and *A Treatise on Wines;* a remarkably brilliant
researcher, he was at the same time a singer of public con-
cert stature.

Thudichum, who worked furiously at his many interests
until his death in 1901, was the founder of brain chemistry.
His determinations of the gross components of the brain
have not, to date, been bested. His skill at unraveling the
chemical structure of many components of the brain is
nothing less than awe inspiring.

But interest in the chemistry of the brain lagged after the
time of Thudichum. The metabolic functions could not vie
for the attention of investigators with the unique functions
of the brain. And so we behold the anomalous situation
that the most remarkable of all organs received the least
attention as an organ. Nevertheless, we learned much about
the brain from borrowed knowledge. Nature is a frugal in-
ventor. If a mechanism works well in one organ, that mechanism is bound to be installed in a new one. Sugars were
broken down and their sun-born energy was sucked up by
cells thousands of years before the specialized cells of the
nervous system made their appearance. When, finally, these
master cells did arise, they were equipped with the very
same furnace for the breakdown of sugars which had been
found efficient in the other cells. The metabolism of sugars
in the brain therefore was unraveled by studying not brain
cells but yeast cells and pigeon liver cells. Thus, borrow-

ing a bit of knowledge here, filling in a bit there, we are slowly rounding out the picture of the brain as a metabolic machine.

Of course, spiritual descendants of Descartes may argue: We grant that the chemist can unravel, slowly, the mechanical functions of the brain; but is there any promise that he can correlate the gross chemical functions and such infinitely subtle entities as thought and personality? Such correlations have been made. Again, these tiny glimpses were made possible not by the work of scientists primarily interested in the brain, but by the work of those whose interests are the vitamins.

Mental disturbances invariably accompany the physical symptoms in pellagra—one of the acute vitamin deficiencies. The patients are apprehensive, fearful, irritable, and easily aroused to anger. These symptoms vanish from pellagrins when nicotinic acid, whose absence is responsible for the disease, is administered.

Nicotinic acid is not unique in throwing the central nervous system out of gear by its absence from a person's diet. In beriberi, induced by the lack of vitamin B_1, severe neurological symptoms are part and parcel of the disease, too. An artificially produced deficiency of biotin converts normal humans into ready subjects for the psychoanalyst's couch.

What is the connection between these vitamins and the functioning of the brain? The source of energy for the brain is the metabolism of sugar. These three vitamins are coenzymes in the various steps of that intricate metabolism. A lack of vitamins is a lack of coenzymes which, in turn, means crippled enzyme systems. It appears that in these cases of vitamin deficiency, deranged enzyme systems in the power plant of the brain produce temporarily deranged personalities.

Whether the stresses on a brain are emotional or chemical (resulting from the absence of dietary essentials), the consequent symptoms are apparently very similar. It may be hasty to jump to the conclusion that the seat of both disorders is necessarily in the same mechanism. (Whether the electric power supply to a radio is erratic or one of its vacuum tubes is worn out, distortion results in both cases.) Nevertheless it is inviting to speculate whether the fragile links in the brain which part under the two different kinds of stresses may not be the same. That the healing is so slow in one case and so spectacular in the other is not a valid objection to tracing the disorders to the same faulty mechanism. Emotional shortcomings are not so easily remedied as biochemical ones.

We know of only one ubiquitous mechanism in living things—enzymes. The "abnormal personality pattern" which makes one person succumb to emotional strains which a sturdier one tosses off, may yet be shown to be an abnormal enzyme pattern. (There is strong evidence that schizophrenics have definite biochemical aberrations from the normal. Under stress, certain enzyme and hormonal systems in such patients are far more sluggish than those of normal persons.)

The time may yet come when we will probe into a patient's erratic enzymes rather than into his emotional history. However, at the present rate of progress, enzyme analysts will not be hanging out their shingles for another couple of hundred years.

Can we approach a means of studying chemically how our brain commands our muscles to move or how that remarkable organ thinks? We have been able to reveal recently a tiny fragment of the chemical machinery which communicates nerve impulses to muscles.

The wiring for the telegraphic system from brain to muscle has been known for a long time. The nerve cells have long, thin threads—some of them several feet long—leading from the spinal cord to the various muscles. Messages are sent along this network—at a speed of about 120 feet per second—ordering the muscles to perform their functions. But there is a gap between the muscle and the end of the nerve fiber; there is no contact at all between the two types of cells. How, then, is a message sent across the gap? What is the messenger which hands over the telegram to the obedient muscle?

Some chemicals, for example, adrenalin, when applied to muscles, make them behave as if they were stimulated by the nerves leading to them. It had been conjectured that perhaps the nerve impulse is conveyed to the muscle by the shooting of some such potent chemical into the space between nerve and muscle.

The conjecture was enshrined as a fact of physiology by a very simple experiment devised by the Austrian pharmacologist Dr. Otto Loewi.[1]

The experiment is remarkable not only for its neat simplicity but also for the circumstances of the birth of the idea: Dr. Loewi dreamed it. He tells that he had a vivid dream in which he performed the crucial experiment. But by the next morning, he had forgotten the details of the fruitful dream. Fortunately, however, he dreamed it again. This time Loewi woke up and jotted down the idea. Then

---

[1] Pharmacology is still another branch of experimental biology. Pharmacologists study the effect of drugs on living things.

he performed the dream-borne experiment which, eventually, also fulfilled a dream: he received the Nobel Prize
in medicine for it.

An organ which, like the heart, operates involuntarily
has two sets of nerves leading to it; one to stimulate it and
one to inhibit or retard it. Thus our brain keeps such an
organ under control with two reins.

Loewi exposed the heart of a frog and severed the nerves
leading to it. He then stimulated with an electric shock the
heart of another frog with intact nerves. The normal result
of such a stimulation is a decrease in the frequency of the
heartbeat. He then withdrew some blood from this stimulated heart and placed it in the first denerved heart. But
nothing happened.

He repeated the very same experiment but he replaced
the blood in the heart having the intact nerves with a salt
solution. This time when he transplanted the salt solution
from the stimulated heart into the nerveless one, something
did happen. The heart which could receive no stimulus
from its severed nerves behaved on receiving the salt solution as if it had been stimulated by a nerve: its heartbeat
decreased. In other words, the stimulated heart had something released into it which could be transferred and was
still potent enough to stimulate another heart. This was the
first proof that there is a messenger which jumps the gap
between the terminal of the nerve telegraphic system and
the addressed muscle cells.

Later, Sir Henry Dale, the British physiologist, proved
the identity of the messenger. It turned out to be a relatively simple chemical called acetylcholine, a substance
which had been made in the laboratory years before. Once
again, the random product of the organic chemist's skill

turned out to be a product of the cell as well. And, also, once again the answer to one question raised other, equally baffling questions: If messages are sent by the release of a chemical, how can the messages be repeated at frequent intervals? Why does not the first message persist through the continued action of the acetylcholine?

The answer is that there is an enzyme present in blood which rips apart acetylcholine into impotent fragments. Thus, once a message is delivered to the muscle cells this enzyme tears up the messenger and the stage is set for a new communication. This is the reason for the failure of Loewi's classic experiment until he replaced the blood in the frog's heart with a salt solution. By the time the blood was transferred from heart to heart, the enzyme had completed the destruction of the messenger. (Incidentally, the German nerve gas DFP acts by inhibiting this very enzyme. Once the enzyme is rendered impotent by DFP, no new messages can be sent to the muscle and the victim becomes paralyzed.)

The work on acetylcholine is testimony to the fact that we can approach with the tools of chemistry and physiology one of the tiny segments of the machinery of the nervous system. But can we at present approach a similar study of the profound functions of the brain? Can we ever answer in chemical terms the ancient plea of Job (Job 28.12): "But where shall wisdom be found and where is the place of understanding?"

Biochemists are confident that some day the molecular site of understanding will be found and that its chemical machinery will be mastered. Unfortunately this confidence springs from the contemplation, not of the current progress in this field, but of history. Many times in the past have scientists, bent on mechanical interpretations of the phenomena of life, encountered forbidding signs: "Thus far and no further." The signs have faded; the barriers between the nonliving and living worlds have crumbled. One by one, every part of life's machinery proved to be physicochemical machinery.

Since we know today that the source of energy for the brain is chemical and that it relays its messages through physicochemical means, it is almost an article of faith that some day we shall find that memory, thought, and will are molecular mechanisms as well.

The metabolism of foods, the release of energy and its conversion into motion, even the propagation of nerve impulses, all this we can explain in chemical terms today. But how does a mass of tissue made of water, some fats, and proteins have a memory? How does it think? How does it write a symphony? Of this we know nothing.

What is memory? Is it carved into the structure of a protein molecule? How is it perpetuated year after year? We know that the brain, like other tissues, is in a state of flux. It is constantly broken down and rebuilt. How does a memory remain intact, sometimes for a lifetime, amidst this constant destruction and repair of tissue? Are the protein molecules which house a memory rebuilt constantly as the antibodies are? Of this, too, we know nothing. This is the depth of our ignorance about the machinery of life.

It is surprising, in view of this impenetrable blanket of ignorance which at present hides completely the mechanism of thought and the shaping of a personality, how ready are some scientists to proclaim at the end of their careers that they have found the ultimate truth in the bottom of their test tubes. In our present state of knowledge, or rather of ignorance, it is presumptuous for scientists to turn into prophets on the basis of their scientific experience. While the tools and methods of science have borne impressive fruits, the area in which those tools can be wielded at present, is quite limited.

Scientists should leave behind their mantles of authority when they abandon the realms explored or explorable by science.

The biochemist can make very thin slices of various organs and can detect enzyme mechanisms in the surviving cells. He can slice liver 0.25 millimeters thick and learn that it makes cholesterol. The butcher slices liver 25 millimeters thick and also knows some of its dietary potencies. The scientist's slices are a hundredfold thinner; his knowledge of the functions of the liver is perhaps a hundredfold more detailed. But is he therefore a hundred times better qualified to ponder the infinity of space, the endlessness of time, the origin of matter-energy?

It is odd how readily a few scientists abandon life-long habits of buttressed reasoning and cautious utterance once they leave their circumscribed fields and take a fling in the wider realms of mysticism. For example, the distinguished physical chemist, the late Pierre Lecomte du Noüy wrote in *The Road to Reason:*

There is an element in the great mystics, the saints, the prophets, whose influence has been felt for centuries, which escapes mere intelligence. We do not admit physical miracles, because they are outside the actual framework of our knowledge; yet we admit the reality of Joan of Arc, who represents a real and confounding miracle.

This is a fallacy of partial truth. Of course Saint Joan is a real and confounding miracle. But so was her lowest yeo-

man a miracle, or, indeed, so was the horse she rode. The
miracle is not a specific life. The miracle is *any* life!

The poet, not unexpectedly, is more sensitive to the mystery and splendor of all things living. Wordsworth wrote:

> To me the meanest flower that blows can give
> Thoughts that do often lie too deep for tears.

There are some scientists who at the end of their careers
enumerate all that is still unknown and, perhaps, unknowa-
ble. On the basis of the enormous gaps in our knowledge
they exhort us to faith. Ignorance of natural phenomena
is an unsteady pillar for the edifice of faith. It is an ephem-
eral stanchion at best. The mystery of yesterday is the com-
monplace of today; the unknown of now will be explored
tomorrow. Three hundred years ago the mechanism of fire
was just as baffling as the workings of the human mind still
are today. (It was thought that substances gave off "phlo-
giston" as they burned. This was a versatile entity possess-
ing either a positive or a negative weight, depending on the
increase or decrease in the weight of the ash after the combustion.) Should men have been exhorted to faith in those
days on the basis of the mystic wonder of a fire?

Pasteur, that greatest of biochemists, who pioneered so
much in the physical realms, pre-empted the role of sci-
entist-mystic as well. He wrote in his speech of acceptance
to the French Academy:

> What is beyond? the human mind actuated by an invincible
> force, will never cease to ask itself: What is beyond? . . . It is
> of no use to answer: Beyond is limitless space, limitless time
> or limitless grandeur; no one understands those words. He
> who proclaims the existence of the Infinite—and none can
> avoid it—accumulates in that affirmation more of the super-
> natural than is to be found in all the miracles of all the re-

ligions; for the notion of the Infinite presents that double character that it forces itself upon us and yet is incomprehensible. When this notion seizes upon our understanding, we can but kneel. . . . I see everywhere the inevitable expression of the Infinite in the world; through it the supernatural is at the bottom of every heart. The idea of God is a form of the idea of the Infinite. As long as the mystery of the Infinite weighs on human thought, temples will be erected for the worship of the Infinite, whether God is called Brahma, Allah, Jehova or Jesus; and on the pavement of those temples, men will be seen kneeling, prostrated, annihilated in the thought of the Infinite.[2]

It seems to the writer that the contemporary mystic scientists have added little to this in substance, or, for that matter, in style.

[2] René Vallery-Radot, *The Life of Pasteur* (Doubleday, 1916), p. 342.

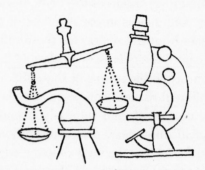

"BIOLOGY began as it will end—as applied chemistry and physics," wrote Dr. Hans Zinsser, the eminent bacteriologist, in *Rats, Lice, and History*.

It is good to have this prophecy from an outstanding classical biologist, for, if it came from a biochemist such a statement might sound like the irresponsible babbling of a partisan enthusiast.

The classical biologist, or rather the biologist with the classical techniques for the study of life, is doomed eventually to technological unemployment. He is doomed by the nature of the object of his study. A living cell is a patterned aggregate of ions and molecules. Life is a sequence of interactions between those ions and molecules. It would appear obvious that the most rewarding study of life must be the study of the physicochemical reactions which *are* life. Unfortunately we are nowhere near such levels of intimacy with the mechanism of life. We have just begun to study life at the molecular level. The road to this stage in the growth of biological science was long and difficult.

The science of gross anatomy, the study of the shape and structure of the organs of the body, began in the sixteenth century. Differences in the structure of the various organs were noted but the reason for those differences remained obscure until the microscope enlarged human vision. With the microscope the source of those differences was found: each different tissue is composed of cells unique to it. During the nineteenth and the early twentieth centuries the anatomical differences discernible under the microscope were gathered into the body of knowledge called histology or microscopic anatomy. At the same time, the functions of these various tissues were studied and were assembled into the discipline called physiology.

Meanwhile the chemist was sharpening his tools and with them was exploring our inorganic universe. Finally, when the defeatist taboos were lifted, he dared to undertake the chemical exploration of the living world as well. It was found that the pancreas dominates the metabolism of sugar not because of the unique structure of that organ but because its cells produce unique molecules, insulin. Thus we arrived at descriptive biochemistry, or, as the author likes to call it, molecular anatomy.

So far the biochemist's efforts have been directed mainly toward the completion of the atlas of molecular anatomy. He has been spectacularly successful in the extraction and identification of the molecular components of the cell. He has catalogued the amino acids, vitamins, hormones, and enzymes. But while all this knowledge is impressive and is of tremendous value in nutrition and in medicine it is astonishingly incomplete knowledge. We know next to nothing about the mode of interaction of those components of the cell. Through the efforts of the biochemist we know the hormone cortisone. That its clinical potency borders on the miraculous is, by now, common knowledge. But how a molecule of that particular structure exerts such a profound influence—of this, we know nothing. We must begin to assemble patiently the basic knowledge for still another science, molecular physiology—the study of the relation between molecular structure and cellular function.

The slightest alteration in the chemical architecture of a component of the cell changes its behavior, altering or abolishing its potency completely. The difference in the structure of linoleic acid, the essential fatty acid whose absence from the diet causes the death of a rat, and oleic acid, which cannot substitute for it, is the deficiency in the linoleic acid of two protons and two electrons. Precisely how the lack of those four minute particles confers life-saving properties on linoleic acid, is not known. To ascribe the differences in the potencies of the two fatty acids to the great specificity of some enzyme system into which the indispensable fatty acid must fit is merely a restatement of ignorance in different terms.

This is an area in which future research promises to bring the richest rewards. For not only do we know little of the mode of action of the simple molecules in the cell, but we know nothing of the relation between structure and function of the huge pivotal molecules of the cell, the proteins. Indeed, here we know little of the structure itself. The riddles of the normal and abnormal functioning of the cell are locked within the so far unknown structure of the protein molecule. The monstrous riddles of the so-called "degenerative diseases"—cancer, multiple sclerosis, arteriosclerosis—are undoubtedly there also.

Through knowledge of descriptive biochemistry the two

other large categories of man's diseases—the deficiency diseases and the infectious diseases—have been mastered. The external deficiencies can be remedied by administering vitamins, essential amino acids, and the appropriate minerals. If the deficiency is an internal one, such as an insufficient supply of hormones, the diminished production can, in many cases, be supplemented by extracts of the organs of other animals or by synthetic preparations. The infectious diseases have been all but mastered with the antimetabolites, the antibiotics, and other chemotherapeutic agents. It is but a question of time before drugs effective against the few remaining recalcitrant infectious agents are found.

Against the degenerative diseases, however, science and medicine are still well nigh powerless. Progress against these, it is feared, must await more extensive knowledge of molecular physiology. This is a realm, the study of the relation between molecular structure and function, where the techniques of the chemist may prove inadequate. Here the physicists will have to be called upon for they are better equipped to study the vagaries of behavior of the ultimate particles of matter and of their aggregates. This is the stage where Zinsser's prophecy will be fulfilled. Biology will become applied physics. However, before that prophecy can be fulfilled there is still time—and need—for several more generations of classical biologists to wield their techniques.

The effective participation by a new discipline in biology must await a definite sequence of development. Most of the fundamental observations in biology have been made by the classical biologists. The biochemists came later to explore the more intimate mechanisms of the phenomena.

It was in 1900 that the laws of heredity discovered by Mendel became generally known. But only in the early 1940s was the chemical pathway of the mechanism of heredity successfully explored: it was found that the genes exert their influence by the production of enzymes. The next problem which awaits solution, probably by the biophysicist, is, how does one speck of matter, the gene, induce the formation of another speck, the enzyme? When that will be answered is impossible to predict.[1]

There is a spectacular, recent example of the contributions physical chemistry can make toward solving a biological and medical problem. The core of a disease known as sickle-cell anemia was traced by a physical chemist to an abnormal protein molecule within the corpuscles of the patients. Sickle-cell anemia is a hereditary disease which apparently afflicts only Negroes. The symptom which distinguishes it from other types of anemia is the presence of abnormally shaped red cells in the veins of the patient: some of the corpuscles are crumpled into crescent or sickle shapes. About 8 percent of our Negro population have some sickle cells in their veins. Fortunately only about 2 or 3 percent have the abnormal cells in sufficiently large numbers to incapacitate them. The puzzling feature of the disease until recently was that the corpuscles have their normal disc shape while they are in the arteries but crumple into shapeless bags in the veins. The disease, there-

[1] Pasteur's work is a contradiction of this ordained sequence of discovery in biology. He was a chemist who made initial observations of biological phenomena and promptly explored them with the tools of chemistry. However, Pasteur, too, made observations of phenomena whose mechanism he could not explore. The current techniques and concepts of chemistry were not up to the task. He could demonstrate that attenuated anthrax bacilli induce immunity to the virulent microorganisms but he did not demonstrate the agents of immunity, the formation of the antibodies. In that instance he was not a biochemist but a biologist.

fore, is not caused merely by a malformation of the cell membranes but rather must be due to an abnormal response either by the membrane or its contents to changes in the concentration of oxygen or of carbon dioxide as the corpuscle exchanges these gases in the lungs.

Dr. Linus Pauling, a former president of the American Chemical Society, is a physical chemist who became interested in biological problems several years ago. He heard of sickle-cell anemia for the first time while he was serving on a committee appointed by President Roosevelt to study means of advancing medicine. Dr. Pauling very naturally approached the disease as a physicochemical problem. He and his associates isolated hemoglobin from the blood of patients with sickle-cell anemia. They then extracted from the hemoglobin the red pigment, heme, in a pure form and compared it with heme obtained from normal humans. The hemes from the two sources were exactly the same. The protein fraction, the globin, was next purified and was made the target of the battery of critical tests and measurements which reveal differences in proteins. One such test is the measurement of the rate of migration of a protein toward one of the poles in an electrical field. Proteins have the ability to assume either a positive or a negative surface charge depending upon the acidity of the liquid in which they are suspended. (Their ambivalence has earned them the name "zwitter-ions" meaning hermaphroditic charged particles.) The electrical charges on the surface of the protein molecule will determine the direction of migration in an electrical field. A level of acidity was reached at which the globin from normal red cells moved toward the positive pole and the globin from sickling cells moved toward the negative pole. In other words the protein from the diseased cells seems to carry a larger positive charge than the normal protein does. Such an abnormal charge can have profound effects on the protein's ability to bind water or the gases carbon dioxide and oxygen to it, and, also, on the ability of the protein molecules themselves to be bound together. Pauling's discovery revealed the cause of the "sickling" phenomenon: when the abnormal protein is laden with carbon dioxide in the veins it shrivels in volume and fails to fill out the cell membrane. His finding also provides a starting point for an approach to a possible remedy of the disease. An attempt to alter the properties of the abnormal protein by means of chemical agents is under way.

Pauling called sickle-cell anemia a "molecular disease." While the name is effective in calling attention to the necessity and fruitfulness of studying biological mechanisms at the molecular level, it is somewhat misleading. All the metabolic diseases and probably all of the degenerative diseases are "molecular." In the case of sickle-cell anemia we happen to know the specific molecule which is "diseased." Other than that the disease is not unique and does not merit a special, generic name.

Could the development of biology be speeded up by inducing large numbers of physicists and physical chemists to become engaged in biological problems? That would hardly speed up the growth of biology any more than putting electricians on the job to start the electrical wiring of a building at the stage of the excavation of its foundation would speed up the completion of the final structure. A biological phenomenon must be recognized, explored, and developed before the physicochemical aspects of it can be seen or approached.

Furthermore, not only must the field be prepared before the physical scientist can be fruitfully engaged in it, but the physical scientists themselves must be prepared. It takes years to train a physicist and then more years of re-training are needed in the handling of biological problems before he is useful to the field, and all of that costs money. Until the Second World War not only was there inadequate financial support for the training of scientists in the United States, but there was precious little support of research al-together. Prior to the war most of the research in the bio-logical sciences was carried on as a part-time, almost a hobby, activity of the professors at the universities and medical schools. Full-time research posts were scarce and in most cases the incumbent in such a post had to have an independent income or had to take a vow of poverty. Furthermore, research programs had to be limited to very modest dimensions because of lack of funds for personnel and equipment. All too often promising leads could not be followed up. The ten-year lag between the discovery and the development of penicillin is perhaps the most vivid example of the tragic neglect of biological research. Flem-ing recognized in 1929 that pencillin may be an "efficient antiseptic" but he failed to concentrate this substance "from lack of sufficient chemical assistance." Had Fleming had a budget in 1929 of perhaps $8,000 a year for his project he could have succeeded with his program of chemical con-centration. But in 1929 it was not easy, even for a distin-guished and seasoned investigator, to obtain $8,000 for a research project in England or in the United States. Obviously, not every promising lead in research will yield rewards as rich as penicillin. But it would have been worth investing in a thousand fruitless searches to bring just one, such as that one, to fruition.

It is not that our economy could not support more research. There was simply a lack of realization of the urgent necessity for such support. It is estimated that prior to 1940 our annual national bill for funeral flowers was one hundred million dollars, an amount which exceeded by far what the Federal government spent on medical research. The scientists themselves were partly to blame for the lack of both appreciation and support of their work. They had secluded themselves in their "ivory laboratories" and refused to communicate with the outside world. There was a time, not too long ago, when a popular article or book by a reputable scientist would cause considerable eyebrow raising. The reporting of the achievements of science was left to individuals from whose writings it appeared that the path to knowledge was illuminated by the flashes of genius of a few eccentric characters in dingy laboratories surrounded by a few pieces of odd glassware and a profusion of disbelieving, jealous colleagues. Recently, the situation has been improved somewhat. There is more extensive dissemination of knowledge by the scientists themselves. There is, also, better reporting of the work of scientists in the newspapers. To be sure, there are some research institutes and individual scientists who have shouldered a little too enthusiastically the burden of communicating their achievements to the general public, for, unfortunately, the knack for self-promotion and the talent for scientific research are not mutually exclusive abilities. This can do but little harm as long as we bear in mind that the headliners of today will not necessarily be the headliners in the history of science, and as long as we leave the evaluation of a scientist to his fellow scientists. Only an expert in the vast literature of science can tell whether an alleged discovery is new, only critical repetition can establish its truth, and time alone can measure its worth.

The support of research, too, has been much improved since the war. Several agencies of the Federal government are financing biological research both in their own institutes and by grants-in-aid to universities. The atom bomb may yet expiate for its existence by bringing the era of neglect of research to an end. For it demonstrated starkly that the research of the "impractical" professors pays and that we must pay for research. (The writer often wonders what would have happened to the atom-bomb project if Congress had had to approve an appropriation of two billion dollars to be put into the hands of professors of physics and chemistry.)

Occasionally in these days there is even a surfeit of enthusiasm to finance research. There was a bill introduced in Congress after the war to appropriate a hundred million dollars—to be spent in one year—for a project on cancer.[2] Fortunately reason (or was it frugality?) prevailed. Spending a hundred million dollars that year would have brought us not one inch closer to the mastery of cancer. Indeed, it is difficult to see how it could have been spent. What is needed is not sudden splurges of funds earmarked to be used to study a specific disease but rather a long-range program of training of scientists and steady support of basic research, as well as campaigns aimed against a specific disease. For if we can base a prediction for the future on the

[2] New York *Times,* February 23, 1946.

experience of the past, mastery of our diseases will come as a natural fruit nourished and ripened on an understanding of the normal mechanisms of the cell. No one whose judgment is worth anything would dare to predict *when,* if ever, the wondrous physicochemical maze bounded by a cell membrane will be sufficiently well explored to make possible the conquest of the degenerative diseases. But one thing we can predict with certainty: without research, biological knowledge—and its applications to medicine—will be at a standstill.

Index